Four
Colors
Suffice

Francis Guthrie (1831–99), originator of the four-color problem.

Four Colors Suffice

How the Map Problem Was Solved

Revised Color Edition

ROBIN WILSON

WITH A NEW FOREWORD BY IAN STEWART

WITHDRAWN

PRINCETON UNIVERSITY PRESS
PRINCETON AND OXFORD

Published by Princeton University Press, 41 William Street, Princeton, New Jersey 08540

In the United Kingdom: Princeton University Press, 6 Oxford Street, Woodstock, Oxfordshire OX20 1TW

First published 2002 by the Penguin Group, Penguin Books Ltd., 80 Strand, London WC2R 0RL

press.princeton.edu

Cover illustration by Eugen Jost.

Cover design by Carmina Alvarez Gaffin.

Library of Congress Cataloging-in-Publication Data

Wilson, Robin J.
 Four colors suffice : how the map problem was solved / Robin Wilson ; with a new foreword by Ian Stewart. — Revised color edition.
     pages cm. — (Princeton science library)
   First published: London : Allen Lane, 2002, under title Four colours suffice.
   Includes bibliographical references and index.
   ISBN-13: 978-0-691-15822-8 (paper : acid-free paper)
   ISBN-10: 0-691-15822-3 (paper : acid-free paper) 1. Four-color problem. 2. Four-color problem—History. 3. Mathematical recreations—History. 4. Proof theory—History. I. Wilson, Robin J. Four colours suffice. II. Title.
   QA612.19.W56 2013
   511'.56--dc23   2013017002

British Library Cataloging-in-Publication Data is available

This book has been composed in Baskerville Pro

Printed on acid-free paper. ∞

Printed in the United States of America

1 2 3 4 5 6 7 8 9 10

IN MEMORY OF JOHN FAUVEL,

WHO ENCOURAGED ME TO WRITE THIS BOOK,

AND OF KENNETH APPEL,

FOR HIS SUPPORT AND ADVICE.

Here of a Sunday morning
  My love and I would lie,
And see the coloured counties,
  And hear the larks so high
  About us in the sky.

A. E. Housman, *A Shropshire Lad*

Suppose there's a brown calf and a big brown dog, and an
artist is making a picture of them . . . He has got to paint
them so you can tell them apart the minute you look at
them, hain't he? Of course. Well then, do you want him
to go and paint both of them brown? Certainly you don't.
He paints one of them blue, and then you can't make no
mistake. It's just the same with maps. That's why they make
every state a different color

Mark Twain, *Tom Sawyer Abroad*

Because the map was printed on a flat surface, only four
colors were required to separate each and every state shape
from its neighbors. On a sphere, a globe, four colors
likewise sufficed. Had the map been printed on a torus—a
doughnut shape—seven colors would have been needed to
allow for the state-shape distinctions. There are, of course,
additional reasons why one seldom encounters a map of the
United States on one's doughnut.

Tom Robbins, *Skinny Legs and All*

CONTENTS

Foreword by Ian Stewart   xi

Preface to the Revised Color Edition   xiii

Preface to the Original Edition   xv

1   The Four-Color Problem   1
What Is the Four-Color Problem? | Why Is It Interesting? |
Is It Important? | What Is Meant by "Solving" It? | Who
Posed It, and How Was It Solved? | Painting by Numbers |
Two Examples

2   The Problem Is Posed   12
De Morgan Writes a Letter | Hotspur and the *Athenaeum* |
Möbius and the Five Princes | Confusion Reigns

3   Euler's Famous Formula   28
Euler Writes a Letter | From Polyhedra to Maps | Only
Five Neighbors | A Counting Formula

4   Cayley Revives the Problem . . .   45
Cayley's Query | Knocking Down Dominoes | Minimal
Criminals | The Six-Color Theorem

5   . . . and Kempe Solves It   55
Sylvester's New Journal | Kempe's Paper | Kempe Chains |
Some Variations | Back to Baltimore

6   A Chapter of Accidents   71
A Challenge for the Bishop | A Visit to Scotland | Cycling
around Polyhedra | A Voyage around the World | Wee
Planetoids

7   **A Bombshell from Durham**   86
Heawood's Map | A Salvage Operation | Coloring
Empires | Maps on Bagels | Picking Up the Pieces

8   **Crossing the Atlantic**   105
Two Fundamental Ideas | Finding Unavoidable Sets |
Finding Reducible Configurations | Coloring Diamonds |
How Many Ways?

9   **A New Dawn Breaks**   124
Bagels and Traffic Cops | Heinrich Heesch | Wolfgang
Haken | Enter the Computer | Coloring Horseshoes

10   **Success!**   139
A Heesch-Haken Partnership? | Kenneth Appel | Getting
Down to Business | The Final Onslaught | A Race against
Time | Aftermath

11   **Is It a Proof?**   157
Cool Reaction | What Is a Proof Today? | Meanwhile . . . |
A New Proof | Into the Next Millennium | The Future

Chronology of Events   171

Notes and References   175

Glossary   187

Picture Credits   193

Index   195

A book about the Four Color Problem, in color. Wonderful! Robin Wilson's *Four Colors Suffice* has been one of my favorite popular mathematics books ever since it was first published, but back then the constraints of the publishing business implied that it had to be printed in black and white. It thereby became a sort of mathematical *Four Shades of Grey*. It was still a terrific read and the mathematical ideas were perfectly clear, but it stands to reason that a book about coloring maps ought to be in color. And now—it is.

I like *Four Colors Suffice* because it manages to tread a very difficult tightrope with consummate ease. It balances the need to entertain with the need to inform. It combines the remarkable and often bizarre history of one of the most celebrated and baffling questions in the whole of mathematics with a masterful presentation of the underlying mathematical ideas. And it achieves this in a way that is accessible to everyone with an interest in mathematics, whatever their level of knowledge might be.

The result is a tale that is readable and gripping, yet manages to possess genuine intellectual depth.

We now know that the answer to the Four Color Problem is indeed "four." This is the smallest number of distinct colors that will color any map drawn on the plane so that regions sharing a common length of boundary have different colors. So now we can call it the Four Color Theorem.

What is so fascinating about this problem is that while anybody can understand the question instantly, and simple experiments with pencils and paper all support the result that everyone thought must be true, it took the world's mathematicians more than a century to answer it. What was needed, to be precise, was a complete, logically impeccable proof that four colors suffice. And that turned out to be extraordinarily hard. Much harder than most mathematicians expected, or imagined.

Right now, all known proofs require serious computer assistance, but as Robin shows us, it wasn't just a simple matter of "putting it on the computer." It never is, in real mathematics, because the interesting problems require a lot more than arithmetic or algebra. They're about structure and ideas, and this one also involves visual reasoning. Often the really hard part, when solving a mathematical problem on a computer, is to massage the question into a form that a computer can work with. That was absolutely the case for this problem.

But let me not anticipate Robin's extraordinary story. I'm happy to include a spoiler that gives away the answer, because Robin does that himself before you even open the book. It's there on the front cover. But I don't want to spoil anything else. I'd rather let Robin take you through it at his own carefully chosen pace and in his own inimitable style.

In color.

Ian Stewart
Coventry 2013

In this revised edition for the Princeton Science series I have made a number of corrections and updated the text to take account of some developments over the past ten years. The book has been redesigned, the spelling has been Americanized, and many of the diagrams now appear in color.

Robin Wilson
April 2013

It is rare for a mathematical problem to catch the attention of the general public. But for a century and a half, the four-color problem on the coloring of maps has been one of the most famous conundrums in the whole of mathematics, if not *the* most famous. Many thousands of puzzlers, amateur problem-solvers, and professional mathematicians have struggled to solve it.

In this book I present the entertaining history of the four-color problem and its solution. It is a story with many interesting and eccentric characters, including Lewis Carroll, the Bishop of London, a professor of French literature, an April Fool hoaxer, a botanist who loved heather, a mathematician with a passion for golf, a man who set his watch just once a year, a bridegroom who spent his honeymoon coloring maps, and a Californian traffic cop.

After defining the problem, I explain the main ideas of the proof, describe the philosophical problems it has raised, and outline some related coloring problems, ranging from the painting of maps on bagels to coloring empires and horseshoes.

For quick reference, a glossary of the technical terms used in the book and a chronology of the main events in the story of the four-color theorem are included after the notes and references that follow the main text.

I should like to thank the following people, who have made valuable material available to me or have improved the book by their comments: Frank Allaire, Ken Appel, Hans-Günther Bigalke, Norman Biggs, Andrew Bowler, Joy Crispin-Wilson, Robert Edwards, Paul Garcia, Wolfgang Haken, Fred Holroyd, John Koch, Stefan McGrath, Donald MacKenzie, Barbara Maenhaut, David Nelson, Susan Oakes, Toby O'Neil, Adrian Rice, Gerhard Ringel, Ted Swart, Stan Wagon, Ian Wanless, Douglas Woodall, and John Woodruff.

The publication of this book coincides with the 150th anniversary of the four-color problem and the 25th anniversary of the publication of its proof.

Robin Wilson
May 2002

Four
Colors
Suffice

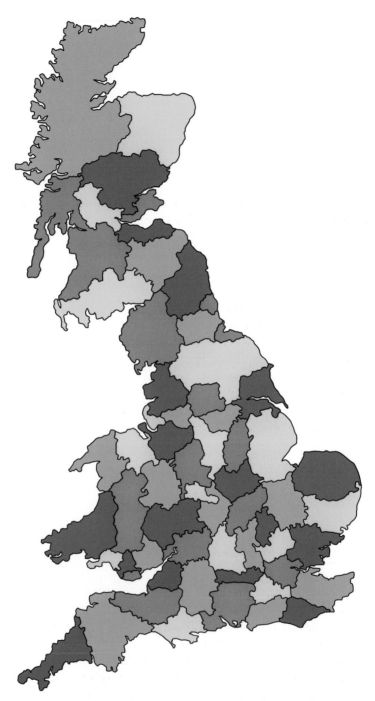

In this map of mainland Britain the counties are colored with four colors in such a way that neighboring counties are colored differently. Can every map be so colored with only four colors?

1

The Four-Color Problem

Before we embark upon our historical journey, there are a number of basic questions to be answered. First and foremost, of course, is this:

What Is the Four-Color Problem?

The *four-color problem* is very simply stated and has to do with the coloring of maps. Naturally, when coloring a map we wish to color neighboring countries differently so that we can tell them apart. So how many colors do we need to color the entire map?

At first sight it seems likely that the more complicated the map, the more colors will be required—but surprisingly this is not so. It seems that at most four colors are needed to color any map—for example, in the map of Britain shown opposite, the counties have been colored with just four colors. This, then, is the four-color problem:

FOUR-COLOR PROBLEM

Can every map be colored with at most four colors in such a way that neighboring countries are colored differently?

Why Is It Interesting?

Solving any type of puzzle, such as a jigsaw or crossword puzzle, can be enjoyed purely for relaxation and recreation, and certainly the four-color problem has provided many hours of enjoyment—and frustration—for many people. At another level, the four-color problem can be regarded as a challenge. Just as climbing a mountain can present a climber with major physical obstacles to overcome, so this problem—so simple to state, yet apparently so hard to solve—presents mathematicians with an intellectual challenge of enormous complexity.

Is It Important?

Rather surprisingly, perhaps, the four-color problem has been of little importance for mapmakers and cartographers. In an article on the problem's origin written in 1965, the mathematical historian Kenneth May observed:

> A sample of atlases in the large collection of the Library of Congress indicates no tendency to minimize the number of colors used. Maps utilizing only four colors are rare, and those that do usually require only three. Books on cartography and the history of mapmaking do not mention the four-color property, though they often discuss various other problems relating to the coloring of maps . . . The four-color conjecture cannot claim either origin or application in cartography.

Equally, there seems to have been no interest shown by quiltmakers, patchworkers, mosaicists, and others in restricting the number of colors of patches or tiles to four.

However, the four-color problem is more than just a curiosity. In spite of its recreational nature, the various attempts to solve it over the years have stimulated the development of much exciting mathematics with many practical applications to important real-world problems. Many practical network problems—on road and rail networks, for example, or communication networks—derive ultimately from map-coloring problems. Indeed, according to a recent book in the related area of graph theory (the study of connections between objects), the entire development of the subject can be traced back to attempts to solve the four-color problem. In the theory of computing, recent investigations into *algorithms* (step-by-step procedures for solving problems) are similarly linked to coloring problems. The four-color problem itself may not be part of the mathematical mainstream, but the advances it has inspired are playing an increasingly important role in the evolution of mathematics.

What Is Meant by "Solving" It?

To "prove the *four-color theorem*" is to show that four colors are sufficient to color all maps—be they geographical maps of the world, or fictitious ones we may care to invent. If its statement were false, we would have to demonstrate the fact by presenting a map that requires five colors or more—just a single map will do. But if the statement is true, then we must prove it for *all possible* maps: it is not enough to color millions or even billions of maps, because there may still be a map that we have missed whose arrangement of countries requires five or more colors. In other areas of science, proofs assert that a given hypothesis is overwhelmingly likely given the underlying assumptions and the experiments that have been carried out, but a mathematical proof has to be *absolute*—no exceptions are allowed. To prove the four-color theorem we must find a general argument that applies to every map, and discovering such an argument turns out to require the development of a great deal of theoretical machinery.

Who Posed It, and How Was It Solved?

As we shall see, the four-color problem was first posed by Francis Guthrie around 160 years ago, but more than a century of coloring maps and developing the necessary theoretical machinery would pass before it was established with certainty that four colors suffice for all maps. Even then, difficult philosophical questions have remained. The eventual solution, by Wolfgang Haken and Kenneth Appel in 1976, required more than a thousand hours of computer time and was greeted with enthusiasm but also with dismay. In particular, mathematicians continue to argue about whether a problem can be considered solved if its solution cannot be checked directly by hand.

PAINTING BY NUMBERS

Before we begin our historical narrative, we need to explain more clearly what the four-color problem is and what our underlying assumptions are. For example, what do we mean by a "map"?

We may think of a *map* as consisting of a number of *countries*, or *regions*. These may be *counties* if part of a British map or *states* if part of a map of the United States. The boundary of each country is made up of several *boundary lines*, and these boundary lines intersect at various *meeting points*. Two countries with a boundary line in common are called *neighboring countries*—for example, in the following map the countries *A* and *B* are neighboring countries.

When coloring a map we must always give neighboring countries different colors. Notice that some maps do need four colors: map (a) below has four mutually neighboring countries, each meeting the other three; these countries must all be colored differently, so four colors are required. This can happen in practice: in map (b), *Belgium*, *France*, *Germany*, and *Luxembourg* are all neighboring countries, so the map requires four colors.

(a)

(b)

Some maps do not need four colors. For example, in each of the following maps, the outer ring of countries can be colored with two colors (*red* and *green*) that alternate as we proceed around the ring; the country in the center must be then assigned another color (*blue*), so that just three colors are required.

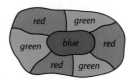

Notice that four colors may be needed even when the map does not contain four mutually neighboring countries. In the map below, the outer ring of five countries cannot be colored with two alternating colors and so requires a third color. The country in the center must then be colored differently from these three, so four colors are required.

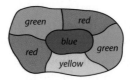

This last situation also arises when we color the forty-eight contiguous states of the United States (excluding Alaska and Hawaii). Here, the western state of *Nevada* is surrounded by a ring of five states, *Oregon*, *Idaho*, *Utah*, *Arizona*, and *California*: this ring of states requires three colors, and *Nevada* must then be assigned a fourth color. This coloring with four colors can then be extended to the entire map.

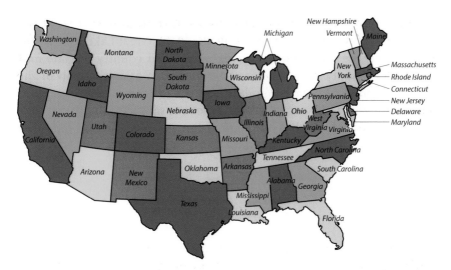

We can make a couple of further observations about this map. Notice first that at one point of the United States four states meet—*Utah, Colorado, New Mexico*, and *Arizona*. We shall adopt the convention that when two countries meet at a single point, we are allowed to color them the same—so *Utah* and *New Mexico* may be colored the same, as may *Colorado* and *Arizona*. This convention is necessary, since otherwise we could construct "pie maps" that require as many colors as we choose—for example, the eight-slice pie map below would need eight colors, since all eight slices meet at the center. With our convention, this map requires only two colors.

Another familiar "map" that needs only two colors is the chessboard. At each meeting point of four squares we alternate the colors *black* and *white*, producing the usual chessboard coloring (see below).

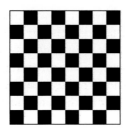

The other feature of the U.S. map we need to note is that the state of *Michigan* is in two parts (separated by one of the Great Lakes) that must be assigned the same color. Although a divided country or state may cause no difficulty in particular situations, such as in the map of the United States, consider the following map in which the two regions numbered 1 are to be regarded as parts of the same country. Here, each of the five "countries" has a boundary line in common with the other four, so five colors are required:

From now on, we avoid such undesirable situations by requiring that each country must be in one piece.

Some people also like to include the *exterior region* in their colorings. Generally speaking, it is immaterial whether we do so or not, since we can regard this exterior region as an extra (ring-shaped) country (see below).

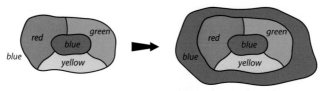

However, a more useful way of thinking about the exterior region is to consider the map as drawn on a globe rather than on a plane. In this case, the exterior region appears as no different from any other region.

In fact, drawing maps on the plane is equivalent to drawing maps on a globe or, in the language of mathematics, on the surface of a sphere.

We can see this from the following illustration, which depicts what is termed a *stereographic projection* from the sphere onto the plane on which it rests. Starting with any map on the sphere, we can project it down from the North Pole to give us a map on the plane. Conversely, given any map on a plane, we can project it up onto a sphere.

Notice that these projections do not affect the coloring of the map—if two neighboring countries are colored *red* and *green*, then after projection (in either direction) their images will also be *red* and *green*. It follows that we can restate the four-color problem as a problem about the coloring of maps on a sphere:

FOUR-COLOR PROBLEM FOR A SPHERE

Can every map drawn on the surface of a sphere be colored with at most four colors in such a way that neighboring countries are colored differently?

If we can solve the four-color problem for maps drawn on a sphere, then we immediately obtain a solution for maps drawn on a plane. Conversely, if we can solve the four-color problem for maps drawn on a plane, then we immediately obtain a solution for maps drawn on a sphere. So it makes no difference whether we think of our maps as drawn on a plane or on the surface of a sphere and so, for each map, we can choose whether or not to color the exterior region.

There are some further simplifications that do not materially affect the four-color problem itself but that will help to simplify our explanation. It almost goes without saying that every map we consider is in one piece, since we could treat the various pieces as maps in their own right and color them all separately. Similarly, we can rule out any map that consists of pieces joined together at a single point, since again we could color the various pieces separately: in particular, we can ignore countries with a single boundary line. Thus, we can ignore any maps like these:

map in two or more pieces   map with pieces joined   one-boundary countries

Finally, we can assume that there are at least three boundary lines at each meeting point. If there were only two boundary lines at a meeting point, then we could simply remove the point without affecting the coloring.

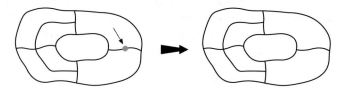

In fact, as we shall see in Chapter 4, when trying to solve the four-color problem we can always restrict our attention to maps where there are *exactly three* boundary lines at each meeting point, as in the map below. Such maps are very common, and we give them a special name: we call them *cubic maps*. You might like to try to color this cubic map with four colors. (Can it be colored with just three?)

TWO EXAMPLES

To end this chapter, here are two practice problems about the coloring of maps on the plane.

Example 1

In the following map, three countries have already been assigned colors. How can we color the entire map with the four colors *red*, *blue*, *green*, and *yellow*?

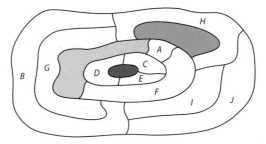

Notice first that country *A* has a boundary line in common with countries colored *green* and *yellow* and so must be colored either *blue* or *red*. Let us look at each of these possibilities in turn.

If country *A* is colored *blue*, then *F* must be colored *red* (since it shares a boundary line with countries colored *blue*, *green*, and *yellow*). Country *D* must then be colored *green*, and *E* must be colored *yellow*. This gives the following coloring.

But it is now impossible to color country *C*, since it has boundary lines in common with countries of all four colors. So *A* cannot be colored *blue*, and must be colored *red*.

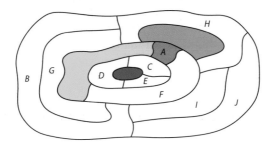

Country *F* must now be colored *blue*, *C* must be colored *green*, *D red*, and *E yellow*. We must then color country *H red*, *G green*, *B yellow*, *I green*, and *J blue*. This completes the coloring.

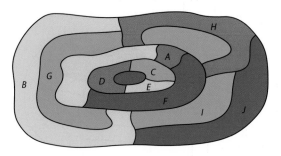

Example 2

Our second example first appeared as an April Fool's joke! For several years Martin Gardner wrote a highly successful mathematical column in the pages of *Scientific American*, and many of his columns were later collected in book form. In the April 1, 1975, issue of the magazine he decided to play a trick on his readers by presenting "Six sensational discoveries that somehow or another have escaped public attention." They included a chess-playing machine (then considered impossible), a thought experiment that disproved Einstein's theory of special relativity, and the discovery of an ancient manuscript establishing Leonardo da Vinci as the first inventor of the valve flush toilet. The column declared:

> The most sensational of last year's discoveries in pure mathematics was surely the finding of a counterexample to the notorious four-color-map conjecture . . . Last November, William McGregor, a graph theorist of Wappingers Falls, N.Y., constructed a map of 110 regions that cannot be colored with fewer than five colors.

In the July 1975 issue the April Fool hoaxer owned up, reporting that his column had elicited more than a thousand letters and that hundreds of readers had sent in colorings of the map with just four colors. McGregor's map appears below. Can *you* color it with four colors?

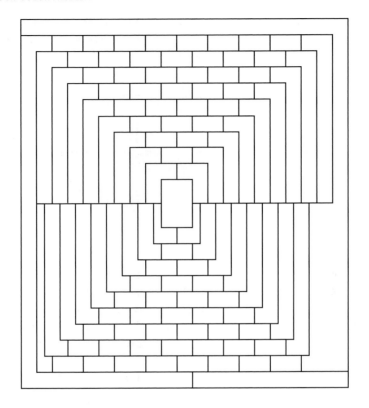

2

The Problem Is Posed

Unlike many problems in mathematics, the origin of the four-color problem can be traced precisely—to a letter written in London in 1852. However, for many years it was believed that the problem could be traced back even further—to a lecture given in Germany around 1840. We start our historical narrative by investigating these rival claims and explaining how the confusion arose.

DE MORGAN WRITES A LETTER

On October 23, 1852, Augustus De Morgan, professor of mathematics at University College, London, wrote to his friend Sir William Rowan Hamilton, the distinguished Irish mathematician and physicist. This was nothing unusual. The two men had corresponded for many years, exchanging family news, reporting on the latest scientific gossip in London and Dublin, and sharing bits of mathematical news. Certainly, neither of them could have imagined that the contents of this particular letter would create mathematical history, for it was here that the *four-color problem* was born.

> A student of mine asked me to day to give him a reason for a fact which I did not know was a fact—and do not yet. He says that if a figure be any how divided and the compartments differently coloured so that figures with any portion of common boundary **line** are differently coloured—four colours may be wanted, but not more—the following is his case in which four **are** wanted

A B C D are names of colors

> Query cannot a necessity for five or more be invented . . .
>
> What do you say? And has it, if true been noticed? My pupil says he guessed it in colouring a map of England . . . The more I think of it

A student of mine asked
me to Day to give him a reason
for a fact which I did not
know was a fact — and do
not yet. He says that. if
a figure be any how divided
and the compartments differently
coloured so that figures with
any portion of common boundary
line are differently coloured
— four colours may be wanted
but not more — The following
is his case in which four
are wanted

A B C &c are
names of
colours

Query cannot a necessity for
five a more be invented

Part of Augustus De Morgan's letter to Sir William Rowan Hamilton, October 23, 1852.

the more evident it seems. If you retort with some very simple case which makes me out a stupid animal, I think I must do as the Sphynx did . . .

Doing as the Sphynx did would have been rather drastic. The Sphynx of ancient mythology was a legendary figure who leapt to her death after Oedipus had correctly solved a difficult riddle she had set him. The riddle was this: What animal walks on four legs in the morning, two at noon, and three in the evening? The answer is man (as a baby, as an adult, and as an elderly person with a stick).

Years later, the student who had approached De Morgan that fateful day identified himself as Frederick Guthrie, subsequently a physics professor and founder of the Physical Society in London. But it was not Frederick who had colored the map of England, as he recalled in 1880:

> Some thirty years ago, when I was attending Professor De Morgan's class, my brother, Francis Guthrie, who had recently ceased to attend them (and who is now professor of mathematics at the South African University, Cape Town), showed me the fact that the greatest necessary number of colours to be used in colouring a map so as to avoid identity of colour in lineally contiguous districts is four. I should not be justified, after this lapse of time, in trying to give his proof, but the critical diagram was as in the margin.

> With my brother's permission I submitted the theorem to Professor De Morgan, who expressed himself very pleased with it; accepted it as new; and, as I am informed by those who subsequently attended his classes, was in the habit of acknowledging whence he had got his information.
>
> If I remember rightly, the proof which my brother gave did not seem altogether satisfactory to himself; but I must refer to him those interested in the subject . . .

Thus it was Frederick Guthrie's elder brother Francis who could justly claim to have originated the four-color problem, but the nature of the "proof" he gave is not known. Francis Guthrie had been a former student of De Morgan's at University College, obtaining a Bachelor of Arts degree there in 1850. Two years later he took a Bachelor of Laws degree and was called to the bar in 1857. He had a distinguished career in South Africa, becoming professor of mathematics at the newly established college at Graaff-Reinet in the Cape Colony, and later at the South African College in Cape Town. A well-liked and popular figure, Guthrie also contributed to botany, which became his chief hobby, and the plant *Guthriea capensis* and the heather *Erica guthriei* were named after him. But he never published anything on the coloring of maps or on the problem that is still sometimes referred to as *Guthrie's problem*.

Frederick Guthrie seems to have been the first to observe that the four-color problem has no interesting extension to three dimensions: if we allow three-dimensional "countries," then we can construct maps that require as many colors as we wish. An example, included by him in the note about his brother, involves a collection of flexible rods (or pieces of colored wool), all touching each other. Since each rod must have a different color from all those it touches, we need as many colors as there are rods: for example, the five rods in Guthrie's diagram below require five colors.

Another three-dimensional example, later described by the Austrian mathematician Heinrich Tietze, involves taking a number of horizontal bars numbered 1 to n, placing on top of them n vertical bars also numbered 1 to n, and then joining (as a single country) each pair of horizontal and vertical bars with the same number. We then obtain n three-dimensional countries, all touching each other, which therefore require n colors. Here, n can be as large as we wish: the following pictures show how to construct the five countries when $n = 5$.

HOTSPUR AND THE *ATHENAEUM*

By 1852 Augustus De Morgan and Sir William Rowan Hamilton were both well established in their respective careers. De Morgan had studied at Cambridge University before becoming the first professor of mathematics at the newly founded University College in London, a position he held for more than thirty years. An eccentric and prolific writer with a style all his own, he is mainly remembered for his popular book *A Budget of Paradoxes*, for De Morgan's laws in set theory, and for his

Augustus De Morgan (1806–1871).

contributions to mathematical logic. Hamilton was a child prodigy, supposedly familiar with Latin, Greek, and Hebrew at the age of 5 and speaking Arabic, Sanskrit, Turkish, and other languages by the time he was 14. He became Astronomer Royal of Ireland while still an undergraduate at Trinity College, Dublin, and held this position until his death in 1865.

As we remarked earlier, De Morgan's 1852 letter to Hamilton was not an isolated event, for they corresponded regularly for thirty years. They met only once, around 1830, when they were introduced to each other by Charles Babbage, whose designs for the so-called analytical engine foreshadowed the invention of the programmable computer a century later. There was a second occasion on which De Morgan and Hamilton were both present, a Freemasons' dinner in honor of the astronomer and mathematician Sir John Herschel, but the event was so crowded that they had no chance to speak to each other.

When De Morgan wrote to Hamilton about the map-color problem, he doubtless hoped that Hamilton would become interested in it. After all, De Morgan had taken an interest in Hamilton's researches, including Sir William's groundbreaking work on *quaternions*. Many mathematical operations are *commutative*, which is to say that they can be carried out in either order: for example, the addition and multiplication of ordinary numbers are commutative ($3 + 4 = 4 + 3$ and $3 \times 4 = 4 \times 3$). However, for Hamilton's quaternions, multiplication is not commutative: his "numbers" are the sum of four terms $a + bi + cj + dk$ (where a, b, c, d are numbers and $i^2 = j^2 = k^2 = -1$) that multiply in a noncommutative way: for example, $i \times j = k$ but $j \times i = -k$, and $k \times i = j$ but $i \times k = -j$.

In the event, De Morgan's letter to Hamilton about the map-color problem drew a terse and idiosyncratic reply: "I am not likely to attempt your "quaternion" of colours very soon." Undeterred, De Morgan wrote to other mathematical friends trying to interest them in the problem. He was fascinated by its intricacies, and in his original letter to Hamilton he had tried to explain where the difficulty lies:

> As far as I see at this moment, if four *ultimate* compartments have each boundary line in common with one of the others, three of them inclose the fourth, and prevent any fifth from connexion with it. If this be true, four colours will colour any possible map without any necessity for colour meeting colour except at a point.
>
> Now, it does seem that drawing three compartments with common boundary A B C two and two—you cannot make a fourth take boundary from all, except inclosing one – But it is tricky work and I am not sure of all convolutions—What do you say? And has it, if true been noticed?

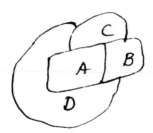

In this passage De Morgan hits upon the fact that if a map contains four regions, each adjoining the other three, then one of them must be completely enclosed by the others. He believed, incorrectly, that this idea lay at the heart of the problem, and it soon became an obsession of his. Since he could not prove it, he proposed to assume its truth as an *axiom*, which he defined as "a proposition which cannot be made dependent upon obviously more simple ones."

In December 1853, De Morgan wrote to the distinguished philosopher William Whewell, Master of Trinity College, Cambridge, describing his observation as a mathematical axiom that had lain "wholly dormant" until it arose in connection with the map-color problem:

I soon made out the following—which was at first incredible—then certainly true—then axiomatic—for I cannot make it depend on anything I see more clearly.

If *four* non-interfering compartments have *each* common boundary *line* with the other three—one at least of the four must be inclosed by the other three—or by fewer . . .

Six months later, in a letter to the Cambridge mathematician Robert Ellis, De Morgan further described it as

an instance of Whewell's views about *latent* axioms, things which at first are not even credible, but which settle down into first principles.

Until recently, the earliest known appearance of the four-color problem in print was also connected with William Whewell. On April 14, 1860, a lengthy unsigned review of Whewell's book *The Philosophy of Discovery, Chapters Historical and Critical* appeared in the *Athenaeum*, a popular literary journal of the time. In his review, the writer outlined the four-color problem, claiming that the problem was familiar to cartographers. He followed his description with a very obscure passage:

Now, it must have been always known to map-colourers that four different colours are enough. Let the counties come cranking in, as Hotspur says, with as many and as odd convolutions as the designer chooses to give them; let them go in and out and round-about in such a manner that it would be quite absurd in the Queen's writ to tell the sheriff that A. B. could run up and down in his bailiwick: still, four colours will be enough to make all requisite distinction.

This mention of Hotspur refers to a passage in Shakespeare's *King Henry IV, Part 1*, where Hotspur remarks, "See how this river comes me cranking in . . ."

The reviewer then asserted that if four areas on a map all have a boundary with the other three, then one area must be surrounded by the others, and this passage clearly identifies De Morgan as the author of the review. De Morgan had indeed written to Whewell on March 3, 1860, thanking him for sending a copy of his book and informing him that he had received a further copy from the *Athenaeum* for review, "which will go back uncut" (in those days one needed a paper cutter to separate the pages of a book).

Recently, an earlier printed reference to the four-color problem has been unearthed. In the *Athenaeum* of June 10, 1854, six years earlier than De Morgan's review, a letter appeared in the *Miscellanea* section:

> *Tinting Maps.*—In tinting maps, it is desirable for the sake of distinctness to use as few colours as possible, and at the same time no two conterminous divisions ought to be tinted the same. Now, I have found by experience that *four* colours are necessary and sufficient for this purpose,—but I cannot prove that this is the case, unless the whole number of divisions does not exceed five. I should like to see (or know where I can find) a general proof of this apparently simple proposition, which I am surprised never to have met with in any mathematical work.   F. G.

It is tempting to ascribe this note to either Francis Guthrie or his younger brother Frederick, but claims have also been made for Francis Galton, a geographer and gentleman of science who was seeking admission to the Athenaeum Club around that time. We shall meet Galton again later."

As a consequence of De Morgan's *Athenaeum* review, the four-color problem crossed the Atlantic to the United States. There it was perused by the American mathematician, philosopher, and logician Charles Sanders Peirce (pronounced "purse"), who developed a lifelong interest in the problem. Peirce considered it "a reproach to logic and to mathematics that no proof had been found of a proposition so simple" and subsequently presented an attempted solution at Harvard University in the presence of his father, Benjamin Peirce, the distinguished Harvard professor of mathematics and natural philosophy. As Charles Peirce later wrote:

> About 1860 De Morgan in the *Athenaeum*, called attention to the fact that this theorem had never been demonstrated; and I soon after offered to a mathematical society at Harvard University a proof of this proposition extending it to other surfaces for which the numbers of colors are greater. My proof was never printed, but Benjamin Peirce, J. E. Oliver, and Chauncey Wright, who were present, discovered no fallacy in it.

In fact, the seminar probably took place in the late 1860s, but the Peirce manuscripts at Harvard University do not indicate the nature of his solution. His reference to "extending it to other surfaces" refers to drawing maps on a surface other than a globe (or sphere). For example, suppose that we lived on a world shaped like the surface of an inner tube or bagel—how many colors would we need then? (Mathematicians call such a surface a *torus*.) From his unpublished notes at Harvard, we know that Peirce found a torus map that needs six colors, but in fact we can do even better than this. The following torus map turns out to have seven mutually neighboring countries and so requires seven colors (we return to the coloring of maps on a torus in Chapter 7):

Peirce later remarked that the four-color problem had been useful to him in testing the growth of his logical powers. Indeed, his researches into mathematical logic included the development of a "logic of relatives," and in October 1869 he specifically applied this to map coloring (his approach is outlined in Chapter 5).

In June 1870, at the beginning of an extensive tour of Europe, Peirce visited the ailing De Morgan in London; it would be fascinating to know whether they discussed the four-color problem. But by this time the problem seems to have been largely forgotten in England: certainly there is no evidence that the recipients of De Morgan's letters were any more interested in the problem than Hamilton had been on first hearing of it in 1852. On March 18, 1871, Augustus De Morgan died in London, having made little progress with the four-color problem and unaware that more than a century would elapse before a solution was discovered.

MÖBIUS AND THE FIVE PRINCES

As we have seen, the four-color problem was originated by Francis Guthrie in 1852. However, it has sometimes been claimed, incorrectly, that the problem is older than this, dating back to a lecture given by the German mathematician and astronomer August Ferdinand Möbius around 1840. The *problem of the five princes* that Möbius posed is superficially similar to the four-color problem, and we shall see how they came to be confused.

August Ferdinand Möbius (1790–1868).

For many years, Möbius was professor of astronomy in Leipzig and director of the Leipzig observatory. In mathematics, his name is associated with the *Möbius function* in the theory of numbers and with *Möbius transformations* in geometry. But he is best remembered for the *Möbius strip*, or *Möbius band*, a curious object constructed from a long rectangular strip of paper by twisting one end through 180° and then gluing the two ends together, as pictured below. The resulting object has just one side and just one edge: this means that an ant could travel from any point on it to any other point without leaving the surface or going over the boundary edge. It was described by the 68-year-old Möbius in late 1858, although it had already been constructed six months earlier by a professor of optics named Johann Benedict Listing, who has received little recognition for his prior discovery.

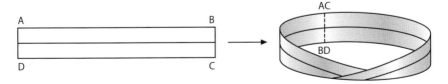

In one of his lectures on geometry, Möbius asked the following question, which had apparently been suggested to him by a Leipzig University friend, the philologist Benjamin Gotthold Weiske, who was greatly interested in mathematics:

PROBLEM OF THE FIVE PRINCES

There was once a king in India who had a large kingdom and five sons. In his last will, the king said that after his death the sons should divide the kingdom among themselves in such a way that the region belonging to each son should have a borderline (not just a point) in common with the remaining four regions. How should the kingdom be divided?

In the next lecture, Möbius's students admitted that they had tried to solve the problem, but without success. Möbius laughed and said he was sorry that they had struggled in vain as such a division of the kingdom is impossible.

It is easy to see intuitively why Möbius's problem has no solution. Suppose that the regions belonging to the first three sons are called A, B, and C. These three regions must all have boundaries in common with one another, as shown below in figure (a). The region D belonging to the fourth son must now lie completely within the area covered by the regions A, B, and C, or completely outside it: these two situations are shown in figures (b) and (c). In each of these situations, it is then impossible to place the region E belonging to the fifth son so as to have boundaries with the other four regions, A, B, C, and D:

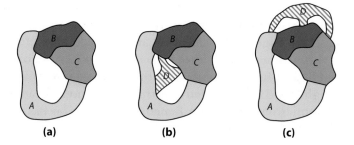

(a)   (b)   (c)

Möbius's problem of the five princes was later extended by Heinrich Tietze, who posed the following related question:

PROBLEM OF THE FIVE PALACES

The king additionally required that each of his five sons should build a palace in his region and that they should link the five palaces in pairs by roads in such a way that no two roads cross. How should the roads be placed?

This problem also has no solution. We can see why by imitating the preceding explanation of the impossibility of solving Möbius's problem.

Suppose that the palaces belonging to the first three sons are called A, B, and C. These three palaces can be linked by noncrossing roads, as shown in figure (a) below. The palace D belonging to the fourth son must now lie completely within the area enclosed by the roads linking A, B, and C, or completely outside it: these two situations are shown in figures (b) and (c). In each of these situations, it is then impossible to build the palace E belonging to the fifth son so as to link it by noncrossing roads to the other four palaces A, B, C, and D.

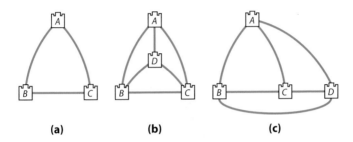

(a)   (b)   (c)

Notice that the solution to either of these problems would have given a solution to the other. If the princes had been able to divide the kingdom into five mutually neighboring regions, then they would also have been able to build palaces in the regions and construct noncrossing roads joining them. On the other hand, if the princes had been able to build the palaces and the roads joining them, then they would have been able to surround these palaces by five neighboring regions. Also, if the king had produced only *four* sons, then the kingdom could easily have been divided, as shown below, and the palaces could have been built and linked by noncrossing roads: with four sons, a solution to either problem yields a solution to the other.

four neighboring regions   superimpose diagrams   four interconnected palaces

Before leaving Möbius's problem of the five princes, we should note that Heinrich Tietze gave a "solution" to it. His description continued:

> The five brothers sank into despair as it became clear that it was not possible to fulfil the condition of their father's will. Suddenly a wandering wizard, who claimed to possess a solution, was announced . . . We can assume that the wizard was richly rewarded.

The wizard's solution was to connect two of the five regions by a bridge:

D and E not connected   D and E joined by a bridge

Of course, this is cheating, since we had assumed that we were re-stricted to drawing the kingdom on a plane, whereas the wizard's solu-tion corresponds to solving Möbius's problem on the surface of a torus.

In fact, for such a problem the king could actually have had up to seven sons: the illustration below shows how to divide a torus into seven neighboring regions, one for each of the seven sons. But if we are restricted to the plane, then, as we have seen, the maximum number of neighboring regions is four—in the plane, five mutually neighboring regions cannot exist.

CONFUSION REIGNS

What is the connection between Möbius's problem of the five princes and the four-color problem, and why were they confused with each other? Before we answer this, let's have a cup of tea and sort out some logic! I can truthfully say that

> if the tea is too hot, then I cannot drink it.

Another way of expressing this is to turn it around, and say that

> if I can drink the tea, then it is not too hot.

What I cannot say is that

> if the tea is not too hot, then I can drink it,

because there may be many other reasons why I cannot drink it: it may be too strong, or too sweet, or have a dead fly in it.

An arithmetical example of this kind of logic, involving the divisibility of whole numbers, is this:

> if a whole number ends with 0, then it is divisible by 5.

For example, the numbers 10, 70, and 530 all end with 0, and all are divisible by 5.

Turning this around, we can deduce that

> if a whole number is not divisible by 5, then it cannot end with 0.

For example, 11, 69, and 534 are not divisible by 5, and none of them ends with 0.

But we cannot say that

> if a whole number is divisible by 5, then it ends with 0,

because there are many numbers, such as 15, 75, and 535, that are divisible by 5 but do not end with 0.

Logicians like to express such statements using symbols. If we use the letter P to stand for "the tea is too hot" or "a whole number ends with 0" and the letter Q to stand for "I cannot drink it" or "it is divisible by 5," we can put the first statement in each of the preceding chains of reasoning into this form:

if P is true, then Q is true—or, equivalently, P implies Q.

Turning this around, we get:

if Q is false, then P is false—or not-Q implies not-P.

But what we cannot say is:

if P is false, then Q is false—or not-P implies not-Q.

Let us now return to Möbius's problem of the five princes. Suppose that the terms of the king's will could be satisfied. Then the region belonging to each of the five sons would share borderlines with the regions belonging to the other four sons—that is, there would be five neighboring regions, each bordering the other four. If we wanted to color these five neighboring regions with different colors, we would need five colors (one for each region). So, the four-color theorem would be false.

The preceding argument tells us that

if there is a map with five neighboring regions,
then the four-color theorem is false.

(Here, P is the statement "there is a map with five neighboring regions," and Q is the statement "the four-color theorem is false." Turning this around, as before, we find that

if the four-color theorem is true,
then there is no map with five neighboring regions.

What we cannot say is that

if there is no map with five neighboring regions,
then the four-color theorem is true.

So our little argument showing the impossibility of solving Möbius's problem does not prove the four-color theorem.

Over the years many people have attempted to prove the four-color theorem by showing that no map can have five neighboring regions. But as we have just seen, this does *not* prove the required result: their logic is the wrong way around.

One unwary person who fell headlong into this trap was the German geometer Richard Baltzer. On January 12, 1885, he gave a lecture to the Leipzig Scientific Society in which he described the problem of the five princes (which he had discovered among Möbius's surviving papers) and then explained why there cannot be five neighboring regions. Baltzer published the results from his lecture, wrongly claiming that the four-color theorem follows immediately from his proof.

Baltzer's published paper was read by Isabel Maddison of Bryn Mawr College, near Philadelphia. In 1897 she wrote a "Note on the history of the map-coloring problem" in the widely read *American Mathematical Monthly*, mentioning Baltzer's paper and remarking that "it does not seem to be generally known that Möbius described the question, in a slightly different form, in his lectures in 1840."

From there, the belief that Möbius was the first to formulate the four-color problem spread far and wide and was given credence when various well-known mathematics books, such as Eric Temple Bell's *The Development of Mathematics*, repeated the error. It was not until 1959 that the geometer H.S.M. Coxeter set the story straight, and since then Francis Guthrie has been universally recognized as the true originator of the four-color problem.

We shall resume our historical development of the four-color problem in Chapter 4, but first we head back to the eighteenth century to investigate the world of polyhedra.

3

Euler's Famous Formula

We begin in Berlin, at the court of Frederick the Great of Prussia. For twenty-five years a notable figure at Frederick's court was the Swiss mathematician Leonhard Euler (pronounced "oiler"): Euler had been appointed by Frederick to the Berlin Academy of Sciences in 1741 and became head of the mathematics division.

At first they got on well—Euler even used to send Frederick strawberries from his garden. However, their relationship quickly deteriorated after the Seven Years' War of 1756–63, when Russian troops occupied Berlin. Frederick began to take an intense interest in the detailed running of the Academy and saw Euler almost every day. Increasingly, Euler came to consider his king as petty and rude, while Frederick, who was a fine composer and performer of music, found the mathematician lacking in social sophistication—rather a bumpkin, in fact. Euler was doubtless relieved to receive an invitation from Empress Catherine the Great in 1766 to return to the St. Petersburg Academy of Science in Russia, where he had been head of the mathematics section before taking up his appointment in Berlin. He spent the remainder of his life in St. Petersburg.

Euler's abilities at mental arithmetic were legendary. When two students tried to sum a complicated progression and disagreed over the fiftieth decimal place, Euler simply calculated the correct value in his head. This led the French physicist François Arago to observe, "He calculated without any apparent effort, just as men breathe, as eagles sustain themselves in the air."

In spite of his increasing blindness, Euler's mathematical output remained high. He was the most prolific mathematician of all time, producing many hundreds of books and papers amounting to tens of thousands of pages, and the publication of his *Collected Works* is still in progress. He wrote on an enormous range of topics, from the purity of prime numbers, musical harmony, and the geometry of the triangle, via calculus and mechanics, to the practical realms of optics, astronomy,

Leonhard Euler (1707–1783).

acoustics, and the sailing of ships. As the applied mathematician Pierre-Simon Laplace would later enthuse to his students,

Read Euler, read Euler. He is the master of us all.

EULER WRITES A LETTER

On November 14, 1750, during his time in Berlin, Leonhard Euler wrote a letter to Christian Goldbach, a former colleague at St. Petersburg and a talented and enthusiastic mathematician. As De Morgan and Hamilton would do a century later, the two men corresponded for many years, sharing news of the latest mathematical developments. Goldbach is now mainly remembered for posing a conjecture that remains unsolved to this day:

every even number greater than 2 can be written as the sum of two prime numbers.

For example,

$$10 = 5 + 5, \quad 20 = 13 + 7, \quad 30 = 19 + 11, \quad \text{and} \quad 40 = 23 + 17.$$

Euler's letter was about *polyhedra*—solid shapes bounded by plane faces, such as a cube (bounded by six squares) or a square pyramid (bounded by a square and four triangles).

The study of polyhedra has a long history, extending at least as far back as the pyramids of ancient Egypt, which date from the third millennium BC. The Greeks were particularly interested in *regular polyhedra*, such as the cube, in which the faces are all regular polygons of the same type (such as squares) and the corners all have the same arrangement of polygons around them. The regular polyhedra, now often called the *Platonic solids*, were discussed by Plato in his dialogue *Timaeus* (ca. 400 BC). Euclid's *Elements* (ca. 300 BC), the most widely read mathematics book of all time, shows how to construct them and concludes with a proof that there can be only five regular polyhedra (see below):

> the *tetrahedron*, bounded by four equilateral triangles;
> the *cube*, or *hexahedron*, bounded by six squares;
> the *octahedron*, bounded by eight equilateral triangles;
> the *dodecahedron*, bounded by twelve regular pentagons;
> the *icosahedron*, bounded by twenty equilateral triangles.

The Greeks associated these polyhedra with the ancient elements: fire (tetrahedron), earth (cube), air (octahedron), and water (icosahedron); the dodecahedron was associated with the cosmos.

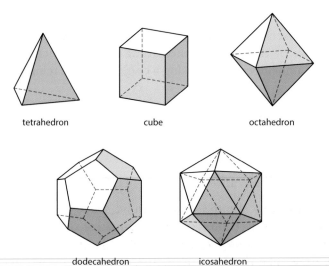

tetrahedron   cube   octahedron

dodecahedron   icosahedron

If we now relax the condition that the faces must all be of the same type but still require the corners to have the same arrangement of regular polygons around them, then we obtain the *semiregular* (or *Archimedean*) *polyhedra*. There are two infinite families of these, the prisms and the antiprisms, consisting of a pair of congruent polygons on the top and bottom, with a strip of squares or equilateral triangles around the middle.

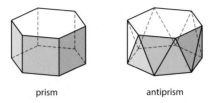

prism   antiprism

There are also thirteen other semiregular polyhedra, some with wonderful names, such as the snub cube and the great rhombicosidodecahedron. Illustrated below are the cuboctahedron (with square and triangular faces), the truncated octahedron (with square and hexagonal faces), the truncated icosahedron (with pentagonal and hexagonal faces), and the great rhombicuboctahedron (with square, hexagonal, and octagonal faces).

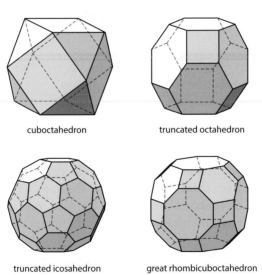

cuboctahedron   truncated octahedron

truncated icosahedron   great rhombicuboctahedron

These polyhedra are not just mathematical curiosities—they are found widely throughout nature: for example, crystals of iron pyrites occur naturally as cubes, octahedra, and dodecahedra, while lead sulphide crystals take the form of cuboctahedra. More recently, certain chemical molecules of carbon atoms, known as *buckminsterfullerenes* and nicknamed *buckyballs*, were found to have the form of polyhedra made up from pentagons and hexagons. (The terms buckminsterfullerene and buckyball come from the name of the architect Buckminster Fuller, whose geodesic domes were based on such polyhedra.) The best-known buckyball has the formula C_{60}, with 60 carbon atoms, and takes the form of a truncated icosahedron, resembling the arrangement of pentagons and hexagons on a soccer ball.

C₃₀ buckyball   C₆₀ buckyball   soccer ball

Although the Greeks and others had been concerned with constructing polyhedra, no one before Euler seems to have had any interest in investigating the possible numbers of faces, edges, and vertices (corners) of a polyhedron. In fact, it was Euler who first introduced the concept of an *edge*, which he called *acies*.

In his letter to Goldbach, Euler wrote, in the curious mixture of Latin and German that he sometimes employed:

Recently it occurred to me to determine the general properties of solids bounded by plane faces, because there is no doubt that general theorems should be found for them . . .

Letting H be the total number of *hedrae* (faces), S the number of *anguli solidi* (vertices, or "solid angles"), and A the number of *acies* (edges), Euler then made a number of assertions. They led to a number of propositions, the sixth of which caused him some difficulty:

But I cannot yet give an entirely satisfactory proof of the following proposition:

6. In every solid enclosed by plane faces the aggregate of the number of faces and the number of solid angles exceeds by two the number of edges, or $H + S = A + 2$ . . .

Euler remarked that "this general result in solid geometry has not previously been noticed by anyone, so far as I am aware." It is now

Part of Euler's letter to Christian Goldbach.

known as *Euler's polyhedron formula*, or *Euler's formula*, for short, and can be expressed as follows:

<u>EULER'S POLYHEDRON FORMULA</u>

For any polyhedron,

(number of faces) + (number of vertices)
= (number of edges) + 2

or, equivalently,

(number of faces) − (number of edges)
+ (number of vertices) = 2.

Rather than using Euler's choice of notation, we shall write F for the number of faces, V for the number of vertices, and E for the number of edges. Euler's polyhedron formula then states that $F + V = E + 2$, or $F - E + V = 2$. We can understand this formula better if we revisit the five regular polyhedra.

For a cube, there are six square faces, twelve edges, and eight vertices, so $F = 6$, $E = 12$, $V = 8$, and Euler's formula yields

$$F - E + V = 6 - 12 + 8 = 2,$$

as expected.

Similarly, for the tetrahedron,

$$F = 4, E = 6, V = 4, \text{ and } F - E + V = 4 - 6 + 4 = 2;$$

for the octahedron,

$$F = 8, E = 12, V = 6, \text{ and } F - E + V = 8 - 12 + 6 = 2;$$

for the dodecahedron,

$$F = 12, E = 30, V = 20, \text{ and } F - E + V = 12 - 30 + 20 = 2;$$

and for the icosahedron,

$$F = 20, E = 30, V = 12, \text{ and } F - E + V = 20 - 30 + 12 = 2.$$

Euler's formula also holds for nonregular polyhedra. For the cuboctahedron,

$$F = 14 \text{ (8 triangles and 6 squares)}, E = 24, V = 12,$$
$$\text{and } F - E + V = 14 - 24 + 12 = 2;$$

for the truncated octahedron,

$$F = 14 \text{ (6 squares and 8 hexagons)}, E = 36, V = 24,$$
$$\text{and } F - E + V = 14 - 36 + 24 = 2;$$

for the truncated icosahedron,

$$F = 32 \text{ (12 pentagons and 20 hexagons)}, E = 90, V = 60,$$
$$\text{and } F - E + V = 32 - 90 + 60 = 2;$$

and for the great rhombicuboctahedron,

$$F = 26 \text{ (12 squares, 8 hexagons and 6 octagons)}, E = 72, V = 48,$$
$$\text{and } F - E + V = 26 - 72 + 48 = 2.$$

Euler's formula has sometimes been wrongly ascribed to the French philosopher René Descartes. Around 1639, Descartes became interested in the total number of angles in all the faces of a polyhedron with F faces and V solid angles (vertices) and he obtained the formula $2F + 2V - 4$. A cube, for example, has six square faces, each with four angles, giving a total of 24 angles, and

$$2F + 2V - 4 = (2 \times 6) + (2 \times 8) - 4 = 24.$$

Descartes also observed that each face has as many angles as sides. But each side is common to two faces, so the total number of angles is equal to $2E$. If we equate these two results we obtain $2F + 2V - 4 = 2E$, from which Euler's formula follows—but Descartes himself never made this link, not least because he did not have the appropriate concepts to do so. *The credit for first stating the polyhedron formula thus belongs to Euler.* Descartes's work on polyhedra is now familiar to us through a copy made by Gottfried Wilhelm Leibniz in the 1670s, but this copy did not come to light until 1859, long after Euler's work on the subject had become well established.

Although Leonhard Euler had initially been unable to "give an entirely satisfactory proof" of his formula, he eventually obtained a "proof by dissection," in which he successively removed tetrahedron-shaped pieces from the polyhedron in such a way that $F - E + V$ remained unchanged at each stage. Continuing in this way, he eventually reached a single tetrahedron, for which $F - E + V = 2$, as we saw before. Since the value of $F - E + V$ remains the same at each stage of the procedure, it must therefore equal 2 for the polyhedron with which he had started. Euler presented his proof to the St. Petersburg Academy of Sciences on September 9, 1751, following it with two substantial papers on polyhedra in the Academy's proceedings, written in 1752 but not published until 1756.

Unfortunately for Euler, it is not at all obvious that his dissection procedure can always be carried out, so his proof must be regarded as deficient. Indeed, the first correct proof was not given until 1794, by the French number theorist, astronomer, and textbook writer Adrien-Marie Legendre. It appeared in his book *Eléments de Géométrie* and used ideas of angle and area that were more akin to Descartes's approach than to Euler's.

FROM POLYHEDRA TO MAPS

What have all these formulas for polyhedra to do with maps? We can see the connection by projecting the polyhedron from a single point onto a plane surface, such as a table top. To do this, we first "inflate" the polyhedron out to a sphere and then project the sphere down onto the plane, as we explained in Chapter 1. The following diagrams show how a cube is projected in this way.

cube   spherical cube   projected cube   cube projection

We can obtain a plane version of any polyhedron in the same way. The following diagrams illustrate a dodecahedron in both its solid and projected versions.

In 1813, Augustin-Louis Cauchy became interested in polyhedra. Cauchy was the leading mathematician in France in the first half of the nineteenth century, obtaining the first rigorous version of the calculus and helping to initiate an important area of algebra now called group theory. He probably has more results and concepts named after him than any other mathematician.

Cauchy proved Euler's formula by first projecting the polyhedron onto a plane surface, as just described. The formula still holds for these plane drawings, provided that we are careful how we deal with the exterior region. Cauchy preferred to ignore this exterior region (which appeared when we flattened the polyhedron), so that he had one region fewer than before. His statement of Euler's formula thus became $F - E + V = 1$; for example, the flattened cube has 5 "interior" regions, 12 edges, and 8 vertices, so $F - E + V = 5 - 12 + 8 = 1$. However, if we include the exterior region, as we usually choose to do when dealing with projections of polyhedra, then $F = 6$, and $F - E + V = 6 - 12 + 8 = 2$, as before.

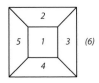

And this is where the connection with maps comes in. We can also think of such plane drawings as maps—the faces correspond to the countries, the edges correspond to the boundary lines, and the vertices correspond to the meeting points.

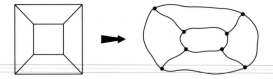

Euler's formula then takes the following form:

EULER'S FORMULA FOR MAPS ON THE PLANE OR SPHERE

If we include the exterior region, then

(no. of countries) − (no. of boundary lines)
+ (no. of meeting points) = 2.

Why is Euler's formula true? One explanation has some features in common with Euler's attempted proof by dissection. We draw a picture of the polyhedron (or map) in the plane and then remove its edges one at a time, always keeping the diagram in one piece, and see what happens to $F - E + V$ at each stage.

How do we remove the edges? There are two possibilities:

1 THE REMOVED EDGE MAY SEPARATE TWO DIFFERENT FACES

Removing such an edge decreases the number of edges by 1, leaves the number of vertices the same, and decreases the number of faces by 1 (since the two faces become merged into one). Thus, E and F are decreased by 1 and V remains unchanged, so $F - E + V$ remains the same.

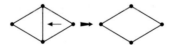

2 THE REMOVED EDGE IS A "HANGING EDGE"

Removing such an edge (and the "hanging vertex" attached to it) decreases the number of edges by 1, decreases the number of vertices by 1, and leaves the number of faces the same. Thus, E and V are decreased by 1 and F remains unchanged, and $F - E + V$ again remains the same.

The following diagrams illustrate this procedure when we successively remove edges from a tetrahedron. Notice how $F - E + V$ remains unchanged throughout the process (in each case, we include the exterior region).

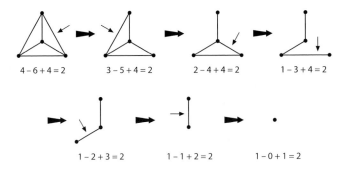

$$4 - 6 + 4 = 2 \qquad 3 - 5 + 4 = 2 \qquad 2 - 4 + 4 = 2 \qquad 1 - 3 + 4 = 2$$

$$1 - 2 + 3 = 2 \qquad 1 - 1 + 2 = 2 \qquad 1 - 0 + 1 = 2$$

Whatever polyhedron or map we start with, we always end up with a single vertex, for which the value of $F - E + V$ is 2. Since this expression remains unchanged throughout the process, we deduce that $F - E + V = 2$ for the original polyhedron. This explains Euler's formula.

Around the same time that Cauchy was publishing his paper, a Swiss mathematician named Simon-Antoine-Jean Lhuilier showed that Euler's formula can fail to hold for polyhedra with certain strange properties. Lhuilier listed three types of strange polyhedra, of which we shall consider just one. If we have a polyhedron with one or more tunnels bored through it, we obtain a version of Euler's formula in which $F - E + V$ is no longer equal to 2. For example, Lhuilier considered a cube with a single tunnel through it and added extra edges so that each face is surrounded by a polygon. The resulting "polyhedron" has 16 faces, 32 edges, and 16 vertices, so that $F - E + V = 16 - 32 + 16 = 0$:

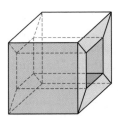

It turns out that $F - E + V = 0$ for any polyhedron with a single tunnel through it. More generally, Lhuilier proved that whenever we bore another tunnel through the solid, we lower the right-hand side of Euler's formula by 2. Thus, for a polyhedron with two tunnels through it, $F - E + V = -2$, and, in general, for a polyhedron with k tunnels, $F - E + V = 2 - 2k$.

In the rest of this chapter we look at two applications of Euler's formula. If you are interested mainly in the historical development of the subject, you can skip the mathematical details, but do note the main results.

ONLY FIVE NEIGHBORS

All the polyhedra we have met in this chapter contain at least one triangle, square, or pentagon, and this is indeed true for all polyhedra. The corresponding result for maps drawn on the plane or sphere is this:

"ONLY FIVE NEIGHBORS" THEOREM

Every map has at least one country with five or fewer neighbors.

This result is very important—indeed, it is central to any proof of the four-color theorem. Since we have excluded one-sided countries, it tells us that every map must contain at least one two-sided country (a *digon*), three-sided country (a *triangle*), four-sided country (a *square*), or five-sided country (a *pentagon*):

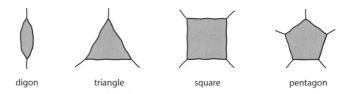

digon   triangle   square   pentagon

To prove the "only five neighbors" theorem, we consider a map with F countries, E boundary lines, and V meeting points and use Euler's formula. From our discussion of maps in Chapter 1 we can assume that in our map there are at least three boundary lines at each meeting point.

Next, we count the boundary lines emerging from all these meeting points. We appear to get at least $3V$ boundary lines in total, because there are at least three boundary lines at each of the V meeting points. But each boundary line has a meeting point at each end and is thus counted twice—so we must divide by 2. So E, the total number of boundary lines, is at least $\frac{3}{2}V$. In symbols, $E \geq \frac{3}{2}V$, which we can turn around and rewrite as $V \leq \frac{2}{3}E$.

We prove that the map has at least one country with five or fewer neighbors by assuming the opposite—that our map has no such country—and showing that this assumption leads to an absurdity. It follows from this assumption that each country is surrounded by at least six neighbors. So, on counting up the boundary lines around all the countries we appear to get at least $6F$ boundary lines in total, be-cause each of the F countries is surrounded by at least six boundary lines. But again, each line has been counted twice, because it has a country on each side, so we must divide by 2. So, E is at least $\frac{6}{2}F$ (which is $3F$). In symbols, $E \geq 3F$, which we can rewrite as $F \leq \frac{1}{3}E$.

We now put these two inequalities $F \leq \frac{1}{3}E$ and $V \leq \frac{2}{3}E$ into Euler's formula, and we get

$$F - E + V \leq \frac{1}{3}E - E + \frac{2}{3}E, \text{ which is } 0.$$

But by Euler's formula, $F - E + V$ is equal to 2, so we have managed to prove the remarkable result that $2 = 0$. This is clearly false. Our error has arisen because we assumed that each country has at least six neigh-bors—so this assumption is unjustified. It follows that at least one coun-try has five or fewer neighbors—which is what we set out to prove.

A COUNTING FORMULA

Our second application of Euler's formula is to obtain a result that has interesting consequences for polyhedra, and which we shall need in Chapter 8—it is called the *counting formula*. For simplicity (and because this is the case that interests us), we limit ourselves to cubic maps, maps in which there are exactly three boundary lines at each meeting point.

Suppose that the map has C_2 two-sided countries (digons), C_3 three-sided countries (triangles), C_4 four-sided countries (squares), and so on. Then F, the total number of countries (including the exterior region), is the sum of all of these numbers:

$$F = C_2 + C_3 + C_4 + C_5 + C_6 + C_7 + \ldots \tag{1}$$

We next count up all the boundary lines in our map in terms of these numbers:

since each digon has two boundary lines, the C_2 digons are surrounded by a total of $2C_2$ lines;

since each triangle has three boundary lines, the C_3 triangles are surrounded by a total of $3C_3$ lines;

since each square has four boundary lines, the C_4 squares are surrounded by a total of $4C_4$ lines;

and so on.

On adding up all these numbers, we appear to get E, the total number of boundary lines. In fact, we get $2E$, twice the total number of boundary lines, because each boundary line lies on the boundaries of two countries—that is,

$$2E = 2C_2 + 3C_3 + 4C_4 + 5C_5 + 6C_6 + 7C_7 + \ldots,$$

which we can rewrite as

$$E = C_2 + \tfrac{3}{2}C_3 + 2C_4 + \tfrac{5}{2}C_5 + 3C_6 + \tfrac{7}{2}C_7 + \ldots \qquad (2)$$

Similarly, we can find the number of meeting points in our map. Since it is a cubic map, there are exactly three boundary lines at each of the V meeting points, giving a total of $3V$ lines altogether. But since each line joins two of the points, we have counted each line twice. So $3V = 2E$, and we can use the preceding expression for $2E$ to get

$$3V = 2C_2 + 3C_3 + 4C_4 + 5C_5 + 6C_6 + 7C_7 + \ldots,$$

which we can rewrite as

$$V = \tfrac{2}{3}C_2 + C_3 + \tfrac{4}{3}C_4 + \tfrac{5}{3}C_5 + 2C_6 + \tfrac{7}{3}C_7 + \ldots \qquad (3)$$

Thus we have expressed F, E, and V in terms of the numbers C_2, C_3, $C_4, \ldots$.

What happens when we substitute the three numbered equations into Euler's formula? We have:

$$2 = F - E + V$$

$$= (C_2 + C_3 + C_4 + C_5 + C_6 + C_7 + \ldots)$$

$$- (C_2 + \tfrac{3}{2}C_3 + 2C_4 + \tfrac{5}{2}C_5 + 3C_6 + \tfrac{7}{2}C_7 + \ldots)$$

$$+ (\tfrac{2}{3}C_2 + C_3 + \tfrac{4}{3}C_4 + \tfrac{5}{3}C_5 + 2C_6 + \tfrac{7}{3}C_7 + \ldots).$$

Rearranging this gives

$$2 = C_2(1 - 1 + \tfrac{2}{3}) + C_3(1 - \tfrac{3}{2} + 1) + C_4(1 - 2 + \tfrac{4}{3}) + C_5(1 - \tfrac{5}{2} + \tfrac{5}{3})$$
$$+ C_6(1 - 3 + 2) + C_7(1 - \tfrac{7}{2} + \tfrac{7}{3}) + \ldots$$
$$= \tfrac{2}{3}C_2 + \tfrac{1}{2}C_3 + \tfrac{1}{3}C_4 + \tfrac{1}{6}C_5 + 0C_6 - \tfrac{1}{6}C_7 - \ldots$$

Finally, on multiplying by 6 to eliminate fractions, we obtain our required counting formula:

COUNTING FORMULA FOR CUBIC MAPS

$$4C_2 + 3C_3 + 2C_4 + C_5 - C_7 - 2C_8 - 3C_9 - \ldots = 12.$$

Notice that the coefficients (the numbers 4, 3, 2, . . . in front of the Cs) successively decrease by 1, and that C_6 (the number of six-sided countries) does not appear since its coefficient is 0.

As a simple illustration of the counting formula, consider this cubic map:

This map has nine four-sided countries and two nine-sided countries (including the exterior one), so $C_4 = 9$ and $C_9 = 2$; all other terms are 0. The counting formula then reduces to

$$2C_4 - 3C_9 = (2 \times 9) - (3 \times 2),$$

which is 12, as expected.

Notice that, in the counting formula, not all the numbers C_2, C_3, C_4, and C_5 can be 0, since otherwise the first four terms on the left-hand side of the equation would disappear, making the left-hand side a negative quantity. But that is impossible, since the right-hand side is 12, a positive number. It follows that at least one of the numbers C_2, C_3, C_4, or C_5 must be positive, and so there must be at least one digon, triangle, square, or pentagon in the map. For cubic maps this gives us an alternative proof of the "only five neighbors" theorem, that every map has at least one country with five or fewer neighbors.

Also, if our map contains no digons, triangles, or squares, the counting formula reduces to

$$C_5 - C_7 - 2C_8 - 3C_9 - 4C_{10} - \ldots = 12.$$

Since the only positive term on the left-hand side is the first one, C_5 must be at least 12. It follows that

every cubic map that contains no digons, triangles, or squares must contain at least twelve pentagons.

We shall use this fact in Chapter 8.

Finally, here are three results on polyhedra that follow from the preceding counting formula for cubic maps. These results hold for any *cubic polyhedron*—a polyhedron in which exactly three faces meet at each vertex.

If all the faces of a cubic polyhedron are pentagons and hexagons, then there are exactly twelve pentagons.

This is because the only nonzero terms are C_5 and C_6, and the counting formula reduces to $C_5 = 12$: an example of such a polyhedron is a *truncated icosahedron*—so all buckyballs and soccer balls must have twelve pentagons!

If all the faces of a cubic polyhedron are squares and hexagons, then there are exactly six squares.

This is because the only nonzero terms are C_4 and C_6, and the counting formula reduces to $2C_4 = 12$, giving $C_4 = 6$: an example of such a polyhedron is a *truncated octahedron*.

*If all the faces of a cubic polyhedron are squares, hexagons, and octagons,
then the number of squares exceeds the number of octagons by 6.*

This is because the only nonzero terms are C_4, C_6, and C_8, and the counting formula reduces to $2C_4 - 2C_8 = 12$, giving $C_4 - C_8 = 6$: an example of such a polyhedron is a *great rhombicuboctahedron*.

4

Cayley Revives the Problem . . .

After Augustus De Morgan's death in 1871, the map-coloring problem lay dormant. Although Charles Sanders Peirce in the United States continued to try to solve it, none of De Morgan's British friends made any mention of it. Yet the problem was not completely forgotten: it certainly continued to languish in the mind of Arthur Cayley of Cambridge University.

Arthur Cayley, a cousin of the inventor and pioneering aviator Sir George Cayley, was an outstanding student at Trinity College, Cambridge. He graduated as Senior Wrangler (best of the year) in 1842 and in October of that year was elected a Fellow of Trinity, the youngest person appointed to a Cambridge Fellowship in the nineteenth century. However, the rules of the university at that time required its fellows to enter holy orders within seven years of receiving their MA degree. Unwilling to do so, Cayley left Cambridge to enter the legal profession, studying at Lincoln's Inn in London.

Cayley was called to the Bar in 1849 and practiced law very successfully for fourteen years. During this period he continued his mathematical researches, publishing no fewer than three hundred papers, many of which contained his best and most original work. In particular, he was one of the founding fathers of matrix algebra and published the first important paper on the subject in 1858. Much of his work on algebra in the early 1850s was with the brilliant but unpredictable James Joseph Sylvester, who had been unable to secure a mathematical position in England and whom we shall meet again in the next chapter.

In 1863 Cayley was elected to the newly founded Sadleirian Chair of Pure Mathematics at Cambridge, obtaining dispensation from the religious requirements. He was elected to an Honorary Fellowship at Trinity College, where he remained for the rest of his life.

Arthur Cayley (1821–1895).

CAYLEY'S QUERY

On June 13, 1878, Arthur Cayley attended a meeting of the London Mathematical Society (LMS) in the rooms of the Royal Asiatic Society at 22 Albemarle Street, Piccadilly. The LMS had been founded at University College, London, in 1865, with Augustus De Morgan as its first president, succeeded in turn by Sylvester and Cayley. By 1878 the number of members had increased from an initial figure of 27 to about 150.

At this particular meeting a former president, Professor Henry Smith of Oxford University, was in the chair, and papers were read on topics ranging from the fluxure of spaces to finding differential resolvents of algebraic equations. As recorded in the Society's *Proceedings*, Cayley then raised a query:

> Questions were asked by Prof. Cayley, F.R.S. (Has a solution been given of the statement that in colouring a map of a country, divided into counties, only four colours are required, so that no two adjacent counties should be painted in the same colour?)

On July 11, Cayley's query was repeated in a report of the meeting in *Nature*.

For almost a century, from then until 1976 when De Morgan's *Athenaeum* review of Whewell's *The Philosophy of Discovery* was located (see Chapter 2), these reports were believed to be the earliest printed references to the four-color problem. One hundred years after that LMS meeting, on June 13, 1978, I commemorated the centenary of Cayley's query with an item about the problem on the BBC's morning radio program *Today*, thereby enhancing breakfast time for several million people. (An earlier suggestion for commemorating the centenary with a television documentary was turned down by the BBC *Horizon* producers as being of "insufficient social relevance"; the following week, *Horizon* broadcast a documentary on the sex life of alligators in South America!)

Cayley was genuinely interested in the four-color problem and was persuaded by his old friend the geographer Francis Galton (whom we met briefly in Chapter 2) to publish an explanatory note on it in the forthcoming new series of the *Proceedings of the Royal Geographical Society*. On receiving Cayley's contribution Galton contacted the editor, recommending publication and saying:

> I wrote to Prof. Cayley of Cambridge, asking if he could send a short paragraph suitable for the Proceedings explaining as far as feasible the very curious problem of the sufficiency of 4 colours to distinguish adjacent territories in maps, which has lately excited great interest among the higher mathematicians.
>
> The problem is especially associated with Prof. Cayley's name & simple as it may be thought at first sight, it still baffles the highest analysis though there is no doubting its truth.
>
> I enclose his reply. It is much longer & more technical than I had hoped, but it makes the character of the problem very clear & I am sure it would interest a certain small section of our society. It is also something to obtain a paper from the most eminent mathematician in England, and probably in the world, on a geographical topic.

Cayley published his note in the April 1879 issue of the *Proceedings*, admitting that

> I have not succeeded in obtaining a general proof: and it is worth while to explain wherein the difficulty consists.

He began by mentioning that the result is "mentioned somewhere by the late Professor de Morgan, who refers to it as a theorem known to map-makers": we now know that the "somewhere" was the *Athenaeum* review of 1860. Cayley then stated the four-color problem and gave the usual four-country example to show that four colors may be needed. He next observed that if a system of an arbitrary number *n* of areas has already been coloured with four colours, and if an (*n* + 1)th area is then added, it by no means follows that we can colour the new area with the same four colours, without altering the original colouring.

For example, he said, if the original coloring has all four colors present on the boundary and if the new area surrounds them all, then there is no color available for the new area and some recoloring is necessary.

coloring of *n* areas   recoloring   coloring of *n* + 1 areas

Cayley then made the useful observation that, when trying to prove that four colors are always sufficient, we can impose more stringent conditions on our maps. In particular, *we can restrict our attention to cubic maps*, ones in which there are exactly three countries at each meeting point. To see why this is true, suppose that there are more than three countries at some meeting points in the map. Over each such point we could stick a small circular "patch", thereby producing a new map in which there are exactly three countries at each meeting point. If we can color this new map with four colors, as assumed, then we easily obtain a coloring of our original map by shrinking each patch down to a point.

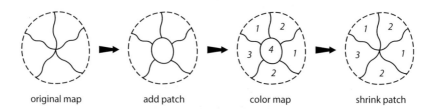

original map   add patch   color map   shrink patch

In a similar way, we can impose other restrictive conditions—for example, as Cayley observed, if we can color all maps with four colors, then we can do so in such a way that only three colors appear on the exterior boundary of the map. This is because we can always add a new ring-shaped country that surrounds the map. Coloring this new map with four colors immediately gives us a coloring of the original map with only three colors on the exterior boundary.

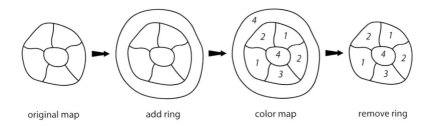

original map   add ring   color map   remove ring

KNOCKING DOWN DOMINOES

Cayley's remark about taking a colored map with n countries, adding an $(n + 1)$th country and then trying to extend the coloring to this new country, leads to a method for tackling the problem. It is known to mathematicians as the *method of mathematical induction*, and it can be traced back at least as far as the fourteenth-century mathematician Levi ben Gerson, inventor of the cross-staff for navigation, who used mathematical induction to prove various results involving permutations and combinations.

In the context of the four-color problem, the method of mathematical induction takes the following form. Suppose that we can prove, for an arbitrary number n, that

if all maps with n countries can be colored with four colors,
then so can all maps with $n + 1$ countries.

We certainly know that all maps with up to four countries can be colored with four colors, and so:

taking $n = 4$, we deduce that all maps with 5 countries can
be colored with four colors;

then, taking $n = 5$, we deduce that all maps with 6 countries
can be colored with four colors;

then, taking $n = 6$, we deduce that all maps with 7 countries
can be colored with four colors;

and so on.

Continuing in this way, we deduce that *all* maps can be colored with four colors.

We can think of a proof by induction as knocking down an unending line of dominoes. We start by knocking down the first domino (or, here, the first four), which in turn knocks down the next one. By our assumption, each domino knocks down the next one—the nth knocks down the $(n+1)$th—so that eventually all the dominoes must fall.

But how can we extend a coloring of a map with n countries to a coloring of a map with $n + 1$ countries? In some instances this is easy: the following coloring extends directly, without any recoloring, and we can color the extra country *red*.

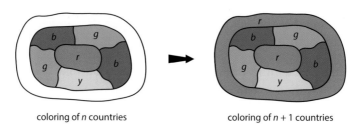

coloring of n countries   coloring of $n + 1$ countries

Sometimes we cannot extend the coloring directly, as happens in the next example, where four different colors appear on the exterior boundary. But if, on the left of the map, we recolor the *red* country *green* and the *green* country *red*, then we can color the extra country *red*.

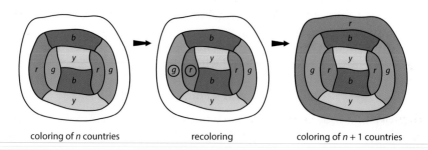

coloring of n countries   recoloring   coloring of $n + 1$ countries

In such simple cases we can extend the given coloring to an extra country without much difficulty, but as our maps become ever more complicated we may need to do a substantial amount of recoloring before we can color the extra country. You might like to try to extend the coloring of the following map to the uncolored country in the middle. (We shall meet this example again in Chapter 7.)

It is clear that a general method for extending a coloring will be hard to find, and therein lies the immense difficulty of the four-color problem.

MINIMAL CRIMINALS

There is an alternative but related approach to solving the four-color problem. Imagine that the four-color theorem is false and that there exist some maps that cannot be colored with four colors. Among these particular maps that need five colors or more, there will be some with the smallest possible number of countries. We call such maps *minimal counterexamples*, or *minimal criminals*—they exhibit "criminal" behavior in that they cannot be colored with four colors, and they are "minimal" because, subject to this condition, they have as few countries as possible. It follows that

a minimal criminal *cannot* be colored with four colors, but any map with fewer countries *can* be colored with four colors.

To prove that four colors suffice, we must prove that minimal criminals cannot exist, and we try to do this by finding ever more restrictive conditions to impose on them. In fact, we try to make life so difficult for them that they cannot exist!

As an example, we can easily show that a minimal criminal cannot contain a two-sided country (a digon). For, suppose that there is a minimal criminal that contains a digon, as shown below. If we remove a boundary line from the digon, merging the digon with one of its neighbors, we obtain a new map with fewer countries. By our assumption, we can color this new map with four colors.

We now reinstate the digon by replacing the removed line. Since four colors are available and since the countries next to the digon use only two of them, there must be a spare color for the digon. Thus, we can color the minimal criminal with four colors, which contradicts our assumption. This shows that a minimal criminal cannot contain a digon.

original map   obtain new map   color new map   color original map

In a similar way, we can show that a minimal criminal cannot contain a three-sided country (a triangle). Suppose that it did, as shown below. We now remove a boundary line, merging the triangle with one of its neighbors. We thus obtain a map with fewer countries, which we can color with four colors, as before.

We now reinstate the triangle. Since the countries next to the triangle use only three of the colors, we can use the fourth to color the triangle. We have thus colored the minimal criminal with four colors, contradicting our assumption. So a minimal criminal cannot contain a triangle.

original map   obtain new map   color new map   color original map

What happens if we try to extend these ideas to a minimal criminal that contains a four-sided country (a square)? As before, we remove a boundary line, merging the square with one of its neighbors and obtaining a map with fewer countries. Again, we can color this new map with four colors.

But when we reinstate the square, the countries next to it may already use all four colors, in which case no spare color is available for coloring the square and the proof cannot proceed, as before.

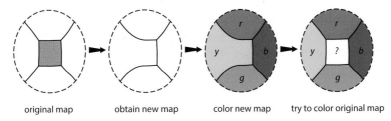

original map   obtain new map   color new map   try to color original map

A similar thing happens if we try to extend these ideas to a minimal criminal that contains a five-sided country (a pentagon). As before, we remove a boundary line, merging the pentagon with one of its neighbors and obtaining a map with fewer countries. As in all the previous cases, this new map can be colored with four colors.

When we reinstate the pentagon, the countries next to the pentagon may already use all four colors, leaving no spare color for the pentagon. Again, the proof cannot proceed.

original map   obtain new map   color new map   try to color original map

In the next chapter we shall see how Alfred Kempe overcame the difficulty when the minimal criminal contains a square, using his *method of Kempe chains*, and then tried to extend his ideas to a minimal criminal that contains a pentagon.

THE SIX-COLOR THEOREM

The idea of a minimal criminal can be used to prove that every map can be colored with six colors. Although rather weaker than the four-color theorem, it is still a remarkable result.

SIX-COLOR THEOREM

> Every map can be colored with six colors in such a way that neighboring countries are colored differently.

To prove the six-color theorem, we first assume that it is false and then look for a contradiction. By our assumption there are maps that cannot be colored with six colors. Among these maps that require seven colors or more, we consider a minimal criminal, one with the smallest possible number of countries. This map *cannot* be colored with six colors, but any map with fewer countries *can* be colored with six colors.

We now use the "only five neighbors" theorem, from Chapter 3. This theorem tells us that our map must contain a country *C* with five or fewer neighbors, as shown below. We now remove a boundary line from the country *C*, merging the country with one of its former neighbors. We then obtain a map with fewer countries, which by our assumption we can color with six colors, *red*, *blue*, *green*, *yellow*, *purple*, and *white*.

original map   obtain new map   color new map   color original map

We now reinstate the country *C*. Since six colors are available and since the countries next to *C* use at most five of them, there must be a spare color for coloring *C*. Thus, we can color the entire map with six colors, which contradicts our assumption. This contradiction shows that no minimal criminal can exist and proves the six-color theorem.

5

. . . and Kempe Solves It

We come now to the most famous fallacious proof in the whole of mathematics—the purported solution of the four-color problem by the London barrister and amateur mathematician Alfred Bray Kempe (pronounced "kemp"). It is unfortunate that he is now remembered mainly for his flawed proof, since he was a fine mathematician, highly regarded by his contemporaries. In Kempe's defence, his error was a subtle one that would remain undetected for eleven years, and his "solution" embodies a number of original ideas that proved to be of the greatest importance in later work on the problem. Kempe's celebrated paper appeared in the *American Journal of Mathematics*, which is where we continue our story.

SYLVESTER'S NEW JOURNAL

Like his mathematical friend Arthur Cayley, with whom he worked so productively in the early 1850s, James Joseph Sylvester had great difficulty in securing an academic position. Until 1871, the professors at Oxford and Cambridge Universities were required to subscribe to the Thirty-nine Articles of the Church of England, and Sylvester, as a Jew, was not eligible for a position in either of the ancient universities. Indeed, though he achieved great success in his degree examinations while studying at St. John's College, Cambridge, he was not allowed to receive the degree until many years later. From 1837 to 1841 he was Professor of Natural Philosophy at the nonsectarian University College, London, where he had briefly studied with Augustus De Morgan years earlier at the age of 14. It was not until 1855, after working for eleven years as an actuary at the Equity & Law Life Assurance Company in London, that he was able to secure a teaching post, at the Royal Military Academy at Woolwich. In 1870, new War Office regulations required all teaching staff at the military academies to retire on half-pay at the age of 55, and Sylvester reluctantly left Woolwich to spend his remaining years in retirement—or so he thought.

James Joseph Sylvester (1814–1897).

For five years Sylvester spent his time publishing a book of poetry, singing in concerts (he reportedly took singing lessons from the French composer Charles Gounod), and dabbling in mathematics. But in late 1875 an event occurred that was to change his life. In America, the new Johns Hopkins University was being founded in Baltimore, Maryland, and its first president, Daniel Gilman, was determined to secure the best faculty members available. Sylvester, by this time possibly the finest mathematician in the English-speaking world, was approached and accepted this new challenge at a salary of $5000 per annum, to be paid in gold.

Sylvester was blissfully happy at Johns Hopkins. He was able to develop his research ideas in a way that had been impossible at Woolwich, where routine teaching was the order of the day. In Baltimore he was surrounded by a group of idealistic and enthusiastic colleagues, such as the mathematician William Story and the astronomer Simon Newcomb, and he set himself the task of stimulating mathematical research activity at Johns Hopkins and, more widely, throughout the United States.

As part of his plans for this ambitious project, Sylvester founded the *American Journal of Mathematics* in 1878, with himself as editor-in-chief and Story as associate editor-in-charge. This publication, which still exists, was designed as a medium of communication between American mathematicians, although "its pages will always be open to contributions from abroad". Indeed, Sylvester commissioned research articles from distinguished friends, both in America and overseas, and the first two volumes contained articles by Arthur Cayley and others from England, as well as papers from France, Germany, and Denmark. Encouraged no doubt by Cayley, the editor-in-chief requested a paper from Alfred Kempe. The paper was entitled "On the geographical problem of the four colours".

KEMPE'S PAPER

Throughout his life, Alfred Bray Kempe was passionately interested in mathematics. At Cambridge he studied under Cayley at Trinity College, graduating in 1872, before embarking on a successful legal career. That same year he wrote his first mathematical paper, on the solution of equations by mechanical means, and five years later, stimulated by Peaucellier's discovery of a linkage for tracing a straight line (illustrated below), he published a popular memoir on linkages entitled "How to draw a straight line".

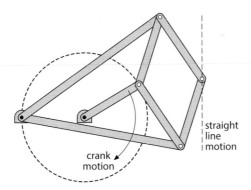

Kempe's work on linkages was so well regarded by his contemporaries that, when he was proposed for a Fellowship of the Royal Society, the first seven of the eight papers cited in support of his proposal were on this subject; the eighth was his paper on the four-color problem. He later became the Royal Society's treasurer, a post he would hold for more than twenty years, and he was knighted in 1912 for his many services to the Society. He was also a keen mountaineer, and

Mount Kempe and the nearby Kempe Glacier in the Antarctic were named after him.

Kempe's interest in map coloring was aroused by Cayley's query at the meeting of the London Mathematical Society—Kempe had attended the meeting—and by Cayley's subsequent memoir in the Royal Geographical Society's *Proceedings* in April 1879. By June of that year, Kempe had obtained his solution of the four-color problem and wrote to Cayley, who replied enthusiastically:

> The solution seems to me quite correct, & I am delighted with it.

On July 17 Kempe published a preview of it in *Nature*, and the full version appeared by the end of the year, in Volume 2 of the *American Journal of Mathematics*. On February 26, 1880, Kempe published simplified versions of his solution in *Nature* and the *Proceedings of the London Mathematical Society*, which corrected some minor errors in his original paper but left intact the fatal flaw.

In his *American Journal of Mathematics* paper, Kempe started by introducing the four-color problem and remarking, as Cayley had done, that De Morgan had "stated somewhere" that the problem was well known to mapmakers. He then mentioned Cayley's contributions and continued:

> Some inkling of the nature of the difficulty of the question, unless its weak point be discovered and attacked, may be derived from the fact that a very small alteration in one part of a map may render it necessary to recolour it throughout. After a somewhat arduous search, I have succeeded, suddenly, as may be expected, in hitting upon the weak point, which proved an easy one to attack . . . How this can be done I will endeavour—at the request of the Editor-in-Chief—to explain.

The rest of the paper falls into three main sections, followed by several shorter observations. In one section Kempe derived an extension of Euler's formula for maps, citing the corresponding plane version derived by Cauchy (see Chapter 3). He stated it thus:

> in every map drawn on a simply connected surface the number of points of concourse [meeting points] and number of districts are together one greater than the number of boundaries

(that is, $V + F = E + 1$). From this, he deduced the formula

$$5d_1 + 4d_2 + 3d_3 + 2d_4 + d_5 - \ldots = 0,$$

Alfred Bray Kempe (1849–1922).

where each d_k denotes the number of regions with k boundaries. This is similar to the counting formula we derived in Chapter 3, except that Kempe also allowed one-sided regions. Since the first five terms are the only positive ones, he deduced that not all the quantities $d_1, d_2, d_3, d_4,$ or d_5 can be zero—in his words, "every map drawn on a simply connected surface must have a district with less than six boundaries." This is the result that we have called the "only five neighbors" theorem.

Using this result, Kempe then described a method for coloring any map. It can be summarized in six steps:

1. Locate a country with five or fewer neighbors (such a country exists, by the preceding result).
2. Cover this country with a blank piece of paper (a *patch*) of the same shape but slightly larger.
3. Extend all the boundaries that meet this patch and join them together at a single point within the patch, as below, which amounts to shrinking the country to a point. It has the effect of reducing the number of countries by one.

4. Repeat this procedure with the new map, continuing until there is just one country left: the whole map is now said to be *patched out*.
5. Color the single remaining country with any of the four colors.
6. Reverse the preceding process, stripping off the patches in reverse order, until the original map is restored. At each stage, color the restored country with any available color until the entire map is colored with four colors.

It was in trying to carry out this final stage that Kempe made his most important contribution to map coloring. For a problem arises: how can we always be sure that one of the four colors is available when a country is restored? As we saw at the end of the previous chapter, there is no difficulty if our restored country has at most three boundary lines. For example, if it is a triangular country, then it is surrounded by three countries that are colored with just three colors; there is then a fourth color available to color the triangle, as shown below. In the language of the previous chapter, this shows that no minimal criminal can contain a triangle.

But what do we do if our restored country has four or five boundary lines—if it is a square or a pentagon? In either case, the restored country may be surrounded by countries that are colored with all four colors, as illustrated below. There is then no color available to color the square or the pentagon.

KEMPE CHAINS

To overcome this difficulty, Kempe introduced a method now known as the *method of Kempe chains*, or a *Kempe-chain argument*. In this method, we look at the colors of the countries surrounding the central one and choose two of them that are not adjacent—say, *red* and *green*. We then look at only those countries that are colored with these colors. We illustrate Kempe's method by first showing how he dealt with a square (which we denote by *S*), surrounded by four countries of different colors: this corresponds to the case where our minimal criminal contains a square.

We look first at the *red* and *green* neighbors of the square *S*. Each of these is the starting point for a *red-green* part of the map—that is, a part of the map consisting entirely of countries colored *red* or *green*. (Although they are called *Kempe chains*, these two-colored parts of the map are not usually chains but may contain "branches," as in the diagrams that follow: these branches may contain arbitrary arrangements of countries. As long as they are colored legitimately, their form has no bearing on the problem of how to color *S*.)

We now ask whether these two *red-green* parts are separate from each other, or whether they link up. Two situations can arise:

case 1   case 2

CASE 1

Here the *red* and *green* countries above *S* that can be reached from the *red* neighbor of *S* do not link up with the *red* and *green* countries below *S* that can be reached from the *green* neighbor of *S*. We can therefore interchange the colors of the *red* and *green* countries above *S*, as shown below, without affecting the coloring of the *red* and *green* countries below *S*. The square *S* is then surrounded only by the colors *green*, *blue*, and *yellow*, so that *S* can be colored *red*. This completes the coloring of the map.

CASE 2

Here the *red-green* part above *S* does link up with the *red-green* part below *S*. This makes things a little more difficult, since nothing is gained by interchanging the colors *red* and *green*: the *red* neighbor of *S* becomes *green*, and the *green* neighbor of *S* becomes *red*, and we are no better off than before.

We turn our attention instead to the *blue* and *yellow* countries and to the *blue-yellow* parts of the map to the left and right of the square *S*. Here, the *blue-yellow* part to the right of *S* is cut off from the *blue-yellow* part to the left of *S*, because the chain of *red* and *green* countries gets in the way.

Thus, we can interchange the colors of the *blue* and *yellow* countries on the right of *S* without affecting the coloring of the *blue* and *yellow* countries on the left of *S*. The square *S* is then surrounded only by the colors *yellow*, *red*, and *green*, so that *S* can be colored *blue*.

This completes the coloring of the map when the restored country is a square and shows that no minimal criminal can contain a square.

Kempe then turned his attention to the case where the restored country is a pentagon (which we denote by *P*), surrounded by five countries that are colored with four different colors. (It is this part of the proof that contains the fundamental flaw, as we shall explain in Chapter 7.)

He again chose two of the surrounding colors that are not adjacent. He first considered the nonadjacent *yellow* and *red* countries above and below *P*. If the *yellow-red* part above *P* does not link up with the *yellow-*

red part below *P*, then we can interchange the colors of the *yellow* and *red* countries above *P* without affecting the coloring of those below *P*.

The pentagon *P* is then surrounded only by the colors *red*, *green*, and *blue*, so that *P* can be colored *yellow*, thereby completing the coloring of the map in this case. We are, therefore, left with the case where the *red-yellow* part above *P* links up with the *red-yellow* part below *P*.

Kempe next considered the nonadjacent *green* and *red* countries above and below *P*. If the *green-red* part above *P* does not link up with the *green-red* part below *P*, then we can interchange the colors of the *green* and *red* countries above *P* without affecting the coloring of those below *P*.

The pentagon *P* is then surrounded only by the colors *red*, *yellow*, and *blue*, so that *P* can be colored *green*, completing the coloring of the map in this case. We are, therefore, left with the case where the *green-red* part above *P* links up with the *green-red* part below *P*. Combining this with the previous case, we obtain:

Notice that the *blue-yellow* part to the right of *P* is separated from the *blue-yellow* part to the left of *P*, because the *red-green* chain of countries gets in the way. We can thus interchange the colors of the *blue-yellow* part on the right of *P* without affecting the coloring of the *blue-yellow* part on the left of *P*.

Similarly, the *blue-green* part to the left of *P* is separated from the *blue-green* part to the right of *P*, because the *red-yellow* chain of countries gets in the way. We can thus interchange the colors of the *blue-green* part on the left of *P* without affecting the coloring of the *blue-green* part on the right of *P*.

If we carry out both of these interchanges, then the pentagon *P* is surrounded only by the colors *yellow*, *red*, and *green*, so *P* can be colored *blue*.

This completes the coloring of the map when the restored country is a pentagon and shows that no minimal criminal can contain a pentagon.

Since we have now dealt with all possible cases, we have proved the four-color theorem.

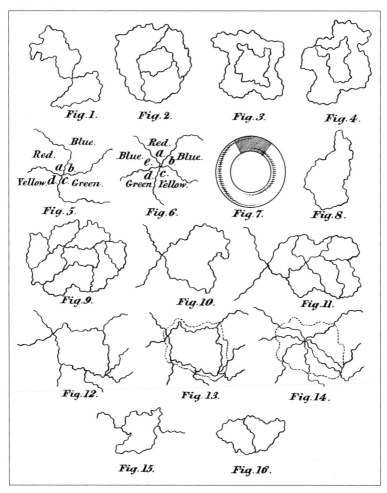

The diagrams from Kempe's paper in the *American Journal of Mathematics*.

SOME VARIATIONS

Although Kempe's proof was incorrect, as we shall explain in Chapter 7, his paper contained a number of further remarks. One of them concerns two special cases of interest that had not been mentioned by previous authors:

If every district of a cubic map is in contact with an even number of others along every circuit formed by its boundaries, three colours will suffice to colour the map.

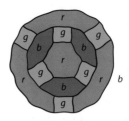

In this case, each country has an even number of neighbors—in particular, the map has no triangles or pentagons—so the countries surrounding any given country must alternate in color.

If an even number of boundaries meet at every point of concourse [meeting point], two colours will suffice.

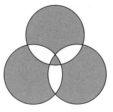

In this case, each meeting point is surrounded by an even number of countries, which must alternate in color. Kempe made the further observation that such a map can be obtained by "drawing any number of continuous lines crossing each other and themselves any number of times": for example, the map above is obtained by drawing three intersecting circles.

A second remark by Kempe is that the ideas of map coloring can be extended to maps drawn on surfaces other than the sphere. Like Charles Sanders Peirce in the 1860s, he remarked that maps drawn on the torus may require six colors—in fact, they may need up to seven colors, as we saw in Chapter 2. This idea of coloring maps on general surfaces would be considerably developed by Percy Heawood in 1890 (see Chapter 7).

Before leaving Kempe's classic paper, we should mention an important connection between map coloring and his work on *linkages*. As Kempe observed:

If we lay a sheet of tracing paper over a map and mark a point on it over each district and connect the points corresponding to districts which have a common boundary, we have on the tracing paper a diagram of a 'linkage', and we have the exact analogue of the question we have been considering, that of lettering the points in the linkage with as few letters as possible, so that no two directly connected points shall be lettered with the same letter.

This construction is similar to the one we used in Chapter 2, when we related the problem of the five palaces to Möbius's problem of the five princes. The linkage obtained from the map of mainland Australia is shown below. Any coloring of the countries of the map gives rise to a lettering of the points in the linkage in which no two directly connected points are lettered the same.

coloring the countries   tracing paper linkage   lettering the points

We now refer to such a linkage as a *graph* (a term coined in 1878 by Sylvester in a chemical context and having no connection with the usual use of that term) and to the preceding process as forming the graph (or dual graph) of the map. This reformulation of the four-color problem as a problem involving the lettering of points reappeared briefly in the 1880s (see Chapter 6) and was later reintroduced in the 1930s and used in all subsequent attempts to solve the problem.

So as not to complicate matters, we shall usually stick to coloring the countries of maps (rather than switching to lettering the points of a graph) throughout the rest of this book.

BACK TO BALTIMORE

In the meantime, Kempe's paper had provoked much interest back at Johns Hopkins University. On November 5, 1879, William Story presented the main ideas of Kempe's proof to an audience of eighteen at a meeting of the Johns Hopkins Scientific Association. Story suggested a few small corrections to Kempe's work, extending the section on Euler's formula to some types of map that Kempe had overlooked. Furthermore, in order "to make the proof absolutely rigorous", he appended a four-page "Note on the preceding paper" to Kempe's paper in the *American Journal of Mathematics*. It is a pity that Story's improvements did not extend to pointing out the fundamental error.

Story's "Note" brought him into dispute with the testy Sylvester who, considering its publication inappropriate and believing his associate editor to have acted unprofessionally, wrote an angry letter to President Gilman:

> When I consider Story's conduct since my absence this year and couple it with the fact of his disobeying my directions when I was absent last year and an inexcusable want of right feeling not to say *mala fides* exhibited by him in his treatment of Mr. Kempe's valuable memoir, I have come to the conclusion that it is inexpedient that we should continue to act together in carrying on the *Journal* . . .

Fortunately, Gilman was able to calm things down, but Story's name ceased to appear on the *Journal*'s title page as Associate Editor in Charge.

Also present at the November gathering was Charles Sanders Peirce, by now a part-time lecturer in logic at Johns Hopkins University, while also working for the U.S. Coastal Survey. Prompted no doubt by Kempe's manuscript, Peirce had written to Story on August 17 mentioning that he had just come across his old solution of the four-color problem—possibly the purported proof that he had presented at Harvard University in the 1860s (see Chapter 2). At the November 5 meeting at which Kempe's proof was discussed, "Remarks were made upon this paper by Mr. C. S. Peirce" and at the following meeting on December 3, Peirce

> discussed a new point in respect to the Geographical Problem of the Four Colours, showing by logical argumentation that a better demonstration of the problem than the one offered by Mr. Kempe is possible.

The details of Peirce's "better demonstration" have not survived, but his recasting of the four-color problem in algebraic terms is outlined below. Here, A, B, C, and D represent colors, and 1, 2, and 3 are countries. A symbol such as C_2 means that color C is assigned to country 2, and he wrote the equation $C_2 = 1$ if country 2 is so colored and $C_2 = 0$ if it is not. Peirce then reformulated the four-color problem as follows (the symbol i refers to any of the countries 1, 2, or 3). An explanation is given immediately after it.

$$A_i^2 - A_i = 0; \text{ and similarly for } B, C, D;$$

$$A_i B_i = 0; \text{ and similarly for all pairs made up of } A, B, C, D.$$

$$\text{Also } A_i + B_i + C_i + D_i = 1.$$

And if i and j are the numbers of two regions having a common boundary, $A_i A_j = 0$, as does the product of any similar pair of Bs, Cs, Ds.

In the first line, the equation $A_i^2 - A_i = 0$ can be solved to give $A_i = 1$ or 0—that is, color A is either assigned to country i or it is not—and similarly for each of the other colors, B, C, and D. The second line asserts that colors A and B cannot both be assigned to country i, and the same is true for any other pair of colors. The third line means that one of the colors A, B, C, or D must be assigned to country i. The fourth line tells us that if the countries i and j have a common boundary line, then color A cannot be assigned to both—and similarly for colors B, C, and D.

Further publicity for Kempe's proof appeared in the pages of the Christmas Day 1879 issue of the New York periodical *The Nation*. Included among its book reviews and descriptions of concerts, plays, and operas was a discussion of the current *American Journal of Mathematics*. This featured an explanation of the "curved ball in baseball" and a "Notice by Mr. Peirce" concerning the first proof of "a mathematical proposition known to be true many years before anyone succeeded in producing a demonstration of it" by "Mr. Kempe who was well-known for his investigation into linkage".

Peirce and Story continued to be interested in the four-color problem for the rest of their lives. Peirce lectured on it at a meeting of the National Academy of Sciences in New York on November 15, 1899, and there are many jottings of maps and their colorings among his various notebooks at Harvard University. We conclude this chapter with an extract from a letter from Story to Peirce that well illustrates the frustrations that the problem caused them.

after Lucas's death, an extended version of it appeared in the final volume of his celebrated four-volume work, *Récréations Mathématiques*.

In Germany, at Richard Baltzer's 1885 Leipzig Scientific Society lecture on Möbius's problem of the five princes (see Chapter 2), the distinguished German mathematician Felix Klein drew Baltzer's attention to "the relevant work of Mr. Kempe in London, on the geographical problem of the four colours". As we have seen, Baltzer confused the two problems, claiming that his proof of the impossibility of five neighboring regions provided a much simpler method for solving the four-color problem and concluding, "How delighted Möbius would have been to see such a far-reaching application of his friend Weiske's economical proof". (Weiske was the colleague who apparently suggested to Möbius the problem of the five princes.) Unfortunately, as we are about to see, Baltzer was not the only person who was accident prone.

A CHALLENGE FOR THE BISHOP

On January 1, 1887, the following paragraphs appeared in the *Journal of Education*. They were sent in by the Rev. James Maurice Wilson, a pioneer in the teaching of science in schools and the Headmaster of Clifton College, a boys' school in the Bristol area.

> One of the 'institutions' at Clifton College is, that the Headmaster sets a Challenge Problem to the whole school once a term. Sometimes these problems refer to mechanical inventions, or applications of electricity; sometimes mathematical problems, which require ingenuity rather than knowledge for their solution. One of his mathematical problems, for example, was to prove that, if all the angular points of a regular decagon are joined, and all the sides and diagonals produced indefinitely, the number of triangles so formed will be 10,000.
>
> He has been good enough to send us the Challenge Problem for last term, which we append. Perhaps it will tempt some of our correspondents to send us a solution.
>
> 'In colouring a plane map of counties, it is of course desired that no two counties which have a common boundary should be coloured alike; and it is found, on trial, that four colours are always sufficient, whatever the shape or number of the counties or areas may be. Required, a good proof of this. Why four? Would it be true if the areas are drawn so as to cover a whole sphere?
>
> 'Solutions to be sent to the Head Master on or before Dec. 1 ... No solution may exceed one page, 30 lines of MS., and one page of diagrams.

The Clifton College challenge problem caused much interest and was taken up in exalted circles. In the June 1, 1889, issue of the *Journal of Education*, Wilson wrote again, saying:

> Some time ago there appeared in your columns a statement of a problem which attracted a good deal of interest . . . Many attempts have been made, and some mathematicians have written to me to say that they could not satisfy themselves with any proof which they had arrived at; and I have been frequently asked whether any neat and satisfactory solutions have been given. I think, therefore, that many of your readers will be glad to see the solution which I now send. It is due to the Bishop of London. He writes:– 'I wrote the enclosed at an evening meeting of the ———— ————, while a vehement ———— was saying what I did not care to listen to.'

Following the meeting, the next morning's papers reported that

> The Bishop of London was greatly interested in Mr. ————'s speech, and was observed to be taking notes all the time he was speaking.

Frederick Temple, Bishop of London and later Archbishop of Canterbury, had once taught mathematics at Balliol College, Oxford, and was particularly interested in mathematical games and puzzles. Unfortunately, his "solution", which was given in full, shows that, like Richard Baltzer and so many others, Temple had considered it sufficient to prove that it is impossible to draw five mutually neighboring regions in the plane. Temple's misunderstanding was explained in detail several years later by John Cook Wilson, Wykeham Professor of Logic at Oxford University.

A VISIT TO SCOTLAND

Another person who was accident prone was Peter Guthrie Tait, professor of natural philosophy in Edinburgh and coauthor, with William Thomson (later Lord Kelvin), of a widely read *Treatise on Natural Philosophy*. Tait was a distinguished mathematical physicist, as well as a passionate golfer, and his interest in the models of trajectories and the behavior of materials under impact led to a classic paper on the trajectory of a golf ball. After carrying out a number of practical experiments, he managed to calculate the greatest theoretical distance that a golf ball can travel and presented his results to the Royal Society of Edinburgh. According to an anecdote related in *Golf* magazine, his son Frederick

Peter Guthrie Tait (1831–1901).

Guthrie Tait, the finest amateur golfer of his time, then proceeded to hit his golf ball five yards further than the greatest distance predicted by his father! However, this report is more likely to have been the product of a journalist's fertile imagination.

Tait was particularly interested in Kempe's note in *Nature* of February 26, 1880, on the solution of the four-color problem. Tait had first learned of the problem from Cayley some years earlier while working on some mathematical problems involving knots, and had independently obtained Kempe's result that if an even number of boundary lines exist at each meeting point in a map, then two colors suffice (though, as he observed, "boundaries usually meet in threes").

Tait considered Kempe's solution of the four-color problem to be unnecessarily lengthy, giving "little insight into its real nature and bearings", and in less than three weeks he produced four or five simpler solutions of his own—none of them correct. These he confidently presented to the Royal Society of Edinburgh on March 15, publishing his results in a paper in its *Proceedings*.

Tait's first abortive approach to the problem used Kempe's linkage diagrams, with points representing countries and lines connecting points corresponding to neighboring countries. He added further lines to divide the diagram into triangles—for, if the points of the "triangulated diagram" can be labeled with four letters so that no two connected points are lettered the same, then (by removing the added lines) so can the points of the original diagram.

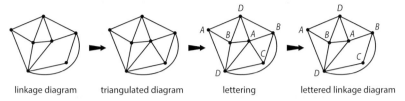

| linkage diagram | triangulated diagram | lettering | lettered linkage diagram |

Tait now added extra points so as to make all the triangular countries four-sided; all the points can then be alternately lettered *A* and *B*. Then, Tait said, "do the same thing again *another way*", giving a second lettering. Finally, he superimposed these two lettered maps, omitting all added points.

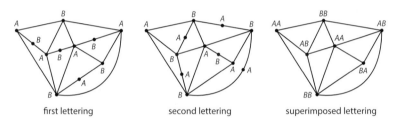

| first lettering | second lettering | superimposed lettering |

Tait claimed that, for every map, this last diagram gives a lettering of the points in the four composite "letters" *AA*, *AB*, *BA*, and *BB*, in which any two connected points are lettered differently. However, it is by no means certain whether the lettering can always be carried out. Tait admitted as much in a letter to Kempe on April 13, 1880, written hastily while he was invigilating an examination:

> There is at first a little puzzle about the two separate ways of lettering the diagram—so as to make the two copies different all over—but it is not difficult to get over that by the help of a few simple rules.

Since he did not explain his "few simple rules" or, indeed, why his method works, his attempt is not a satisfactory solution of the four-color problem.

Later in 1880, Tait presented his one really useful and original idea on the subject, which he believed would yield a solution of the four-color problem. It did not, but it led to an interesting area of study that is still active today. Starting with a cubic map, he considered coloring not its countries but its boundary lines:

> In a map where only three boundaries meet at each point, the boundaries may be coloured with three colours, so that no two of the same colour are coterminous.

(Two boundary lines are *coterminous* if they share a common endpoint.) For example, consider any cubic map whose countries are colored *A*, *B*, *C*, and *D*. We can then color the boundary lines with the three colors α (alpha), β (beta), and γ (gamma), as follows:

> use color α for all lines between pairs of countries colored *A* and *B* or between pairs of countries colored *C* and *D*;
>
> use color β for all lines between pairs of countries colored *A* and *C* or between pairs of countries colored *B* and *D*;
>
> use color γ for all lines between pairs of countries colored *A* and *D* or between pairs of countries colored *B* and *C*.

The procedure is illustrated below:

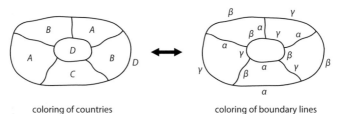

coloring of countries   coloring of boundary lines

This process can always be reversed. Suppose that we are given a cubic map in which the boundary lines have been colored in such a way that all three colors appear at each meeting point. Choose any country and color it with color *A*. The preceding recipe then tells us the colors of the immediately neighboring countries: for example, if the boundary line between the original country and a neighboring one is colored β, then the neighboring country must be colored *C*. Continuing in this way, we can color all the countries of the map.

An alternative way of reversing the process, which will be of importance later, is to choose just two of the colors and look at the boundary lines with these colors: they always form one or more "closed cycles".

map   α and β   α and γ   β and γ

We now take the first of these sets of cycles and write 1 for all countries inside the cycles and 0 for all countries outside. We repeat this for the second set of cycles and superimpose the results:

| map | $a - \beta$ cycles | $a - \gamma$ cycles | superimposed |

This always produces a coloring of the countries in the four "colors" 00, 01, 10, and 11.

Tait correctly realized the importance of coloring the boundary lines of a map and wrote a further letter to Kempe. He considered his earlier result, that the boundary lines of a cubic map can always be colored with three colors, to be

> really the proper thing to start with; as a Lemma easily proved.
>
> So I have withdrawn my long paper and read a much shorter & simpler one (based on the above Lemma) at the last meeting of the R. S. E.
>
> I think I now see the simple principle on wh. the whole depends.

Tait presented this result to the Royal Society of Edinburgh on July 19, 1880, and an abstract duly appeared in the Society's *Proceedings*. Tait believed that his "Lemma easily proved" could be proved easily by mathematical induction (see Chapter 4), and that a solution of the four-color problem could then be deduced from it, as before. Unfortunately, his "Lemma easily proved" is just as difficult to prove as the original four-color theorem.

There is a postscript to Tait's early writings on the four-color problem. His first paper inspired another "Note on the colouring of maps" in the Royal Society of Edinburgh's *Proceedings*. It began:

> From the *Proceedings of the Royal Society of Edinburgh*, No. 106, p. 501, it appears the colouring of maps is receiving attention. This note bears chiefly upon the history of the matter.

The writer was Frederick Guthrie, and this was the note in which he identified his brother Francis as the originator of the four-color problem, as we saw in Chapter 2.

CYCLING AROUND POLYHEDRA

Tait's abstract in the Royal Society of Edinburgh's *Proceedings* was soon expanded into a paper that appeared in the Society's *Transactions*. In this "Note on a theorem in geometry of position", he returned to coloring the boundary lines of a cubic map:

> If $2n$ points be joined by $3n$ lines, so that three lines, and three only, meet at each point, these lines can be divided (usually in many different ways) into three groups of n each, such that one of each group ends at each of the points.

For example, in the following colored diagram with eight points and twelve lines (corresponding to $n = 4$), the lines colored α form one of the three groups, the lines colored β form the second, and the lines colored γ form the third.

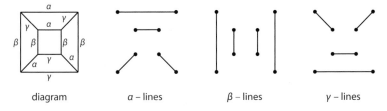

diagram α – lines β – lines γ – lines

Tait then went on to make the confusing assertion that:

> The difficulty of obtaining a simple proof of this theorem originates in the fact that it is not true without limitation.

His example of where the result fails was the following diagram with fourteen points and twenty-one lines (corresponding to $n = 7$). However, this diagram cannot correspond to a map, because the line in the middle does not separate two *different* countries:

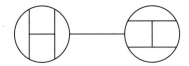

Tait excluded such unwelcome situations by concentrating on cubic maps that can be obtained by projecting a polyhedron onto a plane. For a polyhedron to project to a cubic map it must be a *cubic polyhedron*, such as the tetrahedron, cube, or dodecahedron, in which exactly three edges (or faces) meet at each vertex; the octahedron and icosahedron are not cubic polyhedra, since more than three edges (or faces) appear at each vertex. Tait believed his result to be universally true for all cubic polyhedra, asserting that

> the four-colour theorem implies, and is implied by, the statement that the boundary lines of these polyhedral maps can be coloured with three colours so that all three colours appear at each intersection point.

As a way of attacking this "restricted problem", Tait asked the following question:

> for each cubic polyhedron, is there always a single closed cycle that passes through all the vertices just once?

For the dodecahedron and the truncated octahedron, the following *closed cycles* pass through all the vertices just once:

dodecahedron   truncated octahedron

Tait's interest in these cycles arose from the fact that if a cubic polyhedron has such a cycle, then we can color its edges with three colors by alternating the colors *red* and *green* along the cycle and coloring the remaining edges *blue*. For example, we obtain the following coloring of the edges of a cube:

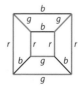

cube   cycle   coloring of edges

In fact, this question of finding a cycle that passes through all the vertices of a polyhedron had been studied more than twenty years earlier by the Reverend Thomas Penyngton Kirkman and Sir William Rowan Hamilton. We now digress briefly and look at their contributions.

A VOYAGE AROUND THE WORLD

Thomas Kirkman was rector of the tiny parish of Croft-with-Southworth, near Warrington, in Lancashire. His parochial duties were not demanding and left him time to have seven children and do a lot of mathematics, becoming a Fellow of the Royal Society in the process. Kirkman was fascinated by polyhedra and tried to find a closed cycle that passes just once through each of the vertices of a given polyhedron. Unfortunately, his papers are not easy to follow since he invented his own terminology, such as a *p-edral q-acron* for a polyhedron with p faces and q vertices and a *triedral summit* (in his spelling) for a vertex at which three faces meet.

Thomas Penyngton Kirkman (1806–1895).

In 1855, Kirkman found a polyhedron that has no such cycle. In his words:

> If we cut in two the cell of a bee by a section of its six parallel edges, we have a 13-acron, whose faces are one hexagon and nine quadrilaterals. The closed 13-gon cannot be drawn.

This polyhedron can be projected onto the plane as follows:

To see why there is no cycle passing through all thirteen vertices, we color the vertices *black* and *white* in such a way that each edge has a *black* end and a *white* end. Any cycle then has to alternate between *black* vertices and *white* ones, and this is possible only if the numbers of *black* and *white* vertices are the same. But there are seven *black* vertices and six *white* ones, so no cycle is possible. However, as Kirkman observed, if we join the central vertex to the one on its left, then we can draw the required cycle.

The following year, William Rowan Hamilton (whom we met briefly in Chapter 2) became interested in drawing cycles on a dodecahedron as a consequence of his work on quaternions—his *icosian calculus*. In particular, he considered three quantities, represented by the Greek symbols ι (iota), κ (kappa), and λ (lambda) that satisfy the equations

$$\iota^2 = 1, \kappa^3 = 1 \text{ and } \lambda^5 = 1, \text{ where } \lambda = \iota\kappa$$

On writing μ (mu) $= \iota\kappa^2$, Hamilton was able to show that $\mu^5 = 1$. He also obtained the lengthy expression

$$\lambda^3\mu^3\lambda\mu\lambda\mu\lambda^3\mu^3\lambda\mu\lambda\mu = 1,$$

which he then proceeded to interpret in terms of cycles on a dodecahedron:

> if we start at the vertex B, proceed towards C, and at each junction think of L as 'turn right' and M as 'turn left', then we obtain the required cycle with the letters in alphabetical order:
>
> B C D F G H J K L M N P Q R S T V W X Z,

ending back at *B*.

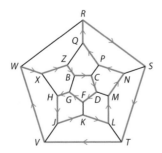

Hamilton was so proud of his icosian calculus that he converted it into *A Voyage Around the World*, a "new and highly amusing game for the drawing room" in which the vertices are labeled with the twenty consonants $B, C, D, \ldots, X, Z$, representing the places Brussels, Canton, Delhi, $\ldots$, Xerez, Zanzibar. The object is to "travel around the world", visiting these cities along the way and returning to one's starting point; one solution is to proceed in alphabetical order, as indicated above. Hamilton "found that some young persons have been much amused" by trying his game and proudly sold it for £25 to the games manufacturer John Jaques and Son of Hatton Garden, London. In the event, this was a wise move on Hamilton's part, as it did not sell.

The *Icosian Game* had twenty numbered pegs that were to be placed into the holes $B, C, D, \ldots$ in cyclic order. The instructions for the game included various puzzles of the form, "given five initial points, in how many ways can the cycle be completed?" For example, how many cycles start with *BCDFG* (there are two), or with *LTSRQ*, or with *JVTSR*?

Such were Sir William's importance and influence that these cycles are now called *Hamiltonian cycles*, rather than being more justly credited to Kirkman, who had beaten Hamilton to them by several months and had also investigated cycles on general polyhedra—not just on a dodecahedron.

WEE PLANETOIDS

We now return to the work of Peter Guthrie Tait. As we saw earlier, Tait was interested in Hamiltonian cycles on cubic polyhedra because they enabled him to color the edges with three colors and thereby to color the faces with four colors. In his November 1880 paper, Tait discussed his "theorem" that every cubic polyhedron has such a cycle and observed that:

> HAMILTON's *Icosian Game* is a particular application of this theorem, the corresponding figure being a projection of a pentagonal dodecahedron. It was suggested to him by the remark, in Mr KIRKMAN's paper on Polyhedra (*Phil. Trans.* 1858, p. 160) that a clear 'circle of edges' of a unique type passed through all the summits of this polyhedron.

Tait was incorrect in saying that Kirkman gave Hamilton the idea for the Icosian Game.

Three years later, Tait followed this paper with a lecture to the Edinburgh Mathematical Society and a further paper, in the *Philosophical Magazine*, in which he reaffirmed his view that all cubic polyhedra have Hamiltonian cycles, remarking:

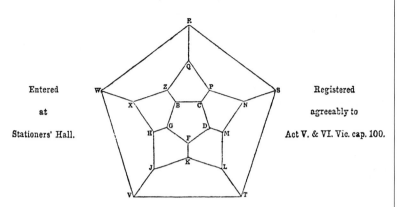

THE ICOSIAN GAME.

Entered

at

Stationers' Hall.

Registered

agreeably to

Act V. & VI. Vic. cap. 100.

In this new Game (invented by Sir WILLIAM ROWAN HAMILTON, LL.D., &c., of Dublin, and by him named *Icosian*, from a Greek word signifying "twenty") a player is to place the whole or part of a set of twenty numbered pieces or men upon the points or in the holes of a board, represented by the diagram above drawn in such a manner as always to proceed *along the lines* of the figure, and also to fulfil certain *other* conditions, which may in various ways be assigned by another player. Ingenuity and skill may thus be exercised in *proposing* as well as in *resolving* problems of the game. For example, the first of the two players may place the first five pieces in any five consecutive holes, and then require the second player to place the remaining fifteen men consecutively in such a manner that the succession may be *cyclical*, that is, so that No. 20 may be adjacent to No. 1; and it is always possible to answer any question of this kind. Thus, if B C D F G be the five given initial points, it is allowed to complete the succession by following the alphabetical order of the twenty consonants, as suggested by the diagram itself; but after placing the piece No. 6 in the hole H, as before, it is *also* allowed (by the supposed conditions) to put No. 7 in X instead of J, and then to conclude with the succession, W R S T V J K L M N P Q Z. Other Examples of Icosian Problems, with solutions of some of them, will be found in the following page.

LONDON:

PUBLISHED AND SOLD WHOLESALE BY JOHN JAQUES AND SON, 102 HATTON GARDEN;

AND TO BE HAD AT MOST OF THE LEADING FANCY REPOSITORIES
THROUGHOUT THE KINGDOM.

The original instructions for Hamilton's Icosian Game.

Probably the proof of this curious proposition has hitherto escaped detection from its sheer simplicity. Habitual stargazers are apt to miss the beauties of the more humble terrestrial objects.

Kirkman had meanwhile returned to the subject of cycles on polyhedra in 1881, perhaps encouraged by Tait's interest in his work. He recast the problem of Hamiltonian cycles on cubic polyhedra in verse form and sent it to the *Mathematical Questions and Solutions of the Educational Times*. Victorian mathematicians frequently expressed their problems in verse, and Kirkman was no exception. Written in the form of a truly appalling piece of doggerel of fifty-three lines, *Question 6610* began:

> 6610. (By the Rev. T. P. KIRKMAN, M.A., F.R.S.)—
> To a wee planetoid, but recently out,
> I am bound to attach an opinion
> On how to effect a design they're about
> Of improving their little dominion.
> Tired of their islands, they long for a Continent:
> Here is the statement they give me, their want anent—

and concluded:

> On islands, three, four, or two,
> Towns, to threescore or two,
> Cover with triedral summits your *n*-edron;
> When they are penned, run
> Over your islands a pencilling cloud,
> Giving the cities the shore-lines allowed.
> The white will all be
> Ferry, cable, and sea.
> You may feel a bit proud,
> If, after some labour, you find what you want, an ent-
> Ire single circle of towns on a Continent.
> When it is found, there is nothing to face,
> But proof of a rule to fit every case.

The "Solution by the PROPOSER" commences with the words:

The theorem—that on every *p*-edron *P*, having only triedral summits, a closed circle of edges passes once through every summit, has this provoking interest, that it mocks alike at doubt and proof.

In other words, it is extremely difficult to decide whether every cubic polyhedron has a Hamiltonian cycle: Kirkman's cell-of-a-bee polyhedron has no Hamiltonian cycle, but it does not constitute a counterexample because it is not a cubic polyhedron.

Kirkman went on to give a plausible proof of the statement, concluding that:

This is the only attempt to prove the general theorem that I desire now to offer. Demonstration it is not; but it will probably carry conviction until a p-edron is produced, on which the sought circle cannot be found. A rigorous demonstration is much to be desired, flowing from the simple definition of a p-edron with triedral summits. But I share the opinion of Professor P. G. TAIT, that our prospect of obtaining such demonstration is very remote indeed. It is my impression that he knows more about these circles than any other.

In the event, it would be a further sixty-five years before the *impossibility* of finding such a proof was demonstrated. In 1946, the cubic polyhedron below was discovered by the English mathematician Bill Tutte.

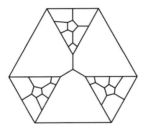

It contains no Hamiltonian cycle, thereby proving the unfortunate Tait wrong yet again.

We conclude this chapter by remarking that, like Kempe, Tutte's name is pronounced with a silent *e*, inspiring a colleague to pen the following words:

Beware the displeasure of Tutte:
He is normally equable, but
He gets in a temper
When people say 'Kemp-e',
And mutters 'Not Tutt-e, but Tutte.'

7

A Bombshell from Durham

Kempe's proof of the four-color theorem was a very good one. It was incorrect, but it was a very good incorrect proof. Not only did it convince Victorian mathematicians for eleven years, but most of the ideas it contained were sound and would form the basis for later assaults on the problem, including the one that was ultimately to prove successful. The credit for finding the error in Kempe's paper goes to Percy Heawood (pronounced "haywood"), a mathematics lecturer at the Durham Colleges (now Durham University).

HEAWOOD'S MAP

Percy John Heawood studied at Exeter College, Oxford, gaining degrees in both mathematics and classics and winning the University's junior and senior mathematics prizes. In 1887 he was appointed mathematical lecturer at Durham, where he spent the rest of his life, becoming professor and eventually vice-chancellor and retiring at the age of 78 in 1939.

Throughout his life, Heawood had a passion for committees and considered a day as wasted if he did not attend at least one committee meeting. A noted Latin, Greek, and Hebrew scholar, he continued with his studies in mathematics and the classics after his retirement, and in his ninetieth year published two research papers—one on the four-color problem in the *Proceedings of the London Mathematical Society*, the other on the date of the Last Supper in the *Jewish Quarterly Review*.

Like several of the other characters in our story, Heawood was somewhat eccentric. As his London Mathematical Society obituary notice affectionately observed:

> In his appearance, manners and habits of thought, Heawood was an extravagantly unusual man. He had an immense moustache and a meagre, slightly stooping figure. He usually wore an Inverness cape of strange pattern and manifest antiquity, and carried an ancient handbag. His walk was delicate and

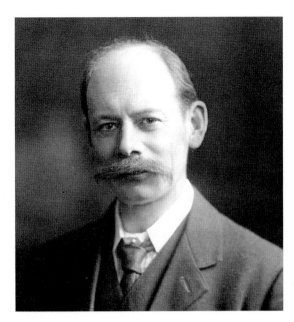

Percy John Heawood (1861–1955).

hasty, and he was often accompanied by a dog, which was admitted to his lectures.

One particularly unusual trait of "Pussy" Heawood (so nicknamed because his moustache resembled a cat's whiskers) was that he used to set his watch just once a year, on Christmas Day. Whenever he needed to know the time, he would do the necessary calculations in his head, based on how quickly it lost time. He reportedly once confided to a colleague who had asked him the time, "No, it's not two hours fast; it's ten hours slow!"

In addition to his work on map coloring, which is described next, Heawood's major achievement was to save the magnificent eleventh-century Durham Castle, set on a promontory above the River Wear, from sliding down the cliff into the water below. In 1928 its foundations were found to be insecure, and money was hard to find. Almost single-handedly, as secretary of the Castle Preservation Committee, Heawood managed to raise the necessary funding and the castle was saved. For these efforts Durham University awarded him an honorary Doctorate of Civil Laws in 1931, and he was created an Officer of the Order of the British Empire (OBE) in 1939.

Percy Heawood's interest in the four-color problem dated back to his first term at Oxford University. In a letter to the Belgian mathematician Alfred Errera in the 1920s, he recalled that:

> When I went up as a student to Oxford in 1880, H.J.S. Smith was the Professor of Geometry. A very clear and interesting lecturer, before proceeding with the proper subject material of his course he started by dividing geometrical properties into (1) properties of situation, (2) descriptive properties, (3) metrical properties; and among two or three examples of (1), he cited the four colour theorem as probably true but unproved; since then, this problem has always fascinated me.

We saw in Chapter 4 that Henry Smith had presided over the London Mathematical Society meeting at which Cayley raised his query about map coloring, but Smith was evidently unaware of Kempe's subsequent solution. Heawood's letter continued:

> I thus turned naturally, when I heard about it, to the so-called proof of Kempe—and after a critical examination, I discovered the error I indicated in my printed note. I had never heard of any doubts raised as to its validity before my paper in the 'Quarterly Journal'.

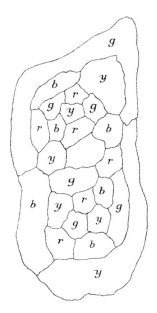

Heawood's counterexample to Kempe's proof.

Heawood wrote his famous paper, "Map-colour theorem", in June 1889, and it duly appeared in the *Quarterly Journal of Mathematics* in June of the following year. In his introduction, Heawood seemed almost apologetic about uncovering the error in Kempe's paper:

> The present article does not profess to give a proof of this original Theorem; in fact its aims are so far rather destructive than constructive, for it will be shown that there is a defect in the now apparently recognized proof . . . .

Heawood's demolition of Kempe's proof centers on the argument that Kempe used when dealing with a pentagon. As we saw in Chapter 5, Kempe considered an arrangement of countries like this:

(Here the colors surrounding the pentagon have been changed to agree with the example in this chapter.) Kempe then used two simultaneous interchanges of color to recolor the countries on either side of the pentagon so that the pentagon itself could then be colored. Either interchange of color is perfectly valid, but to do both at once is not permissible.

To explain why this is so, Heawood introduced the map opposite. In this map, each of the twenty-five countries has been colored *red, blue, yellow,* or *green,* except for the central pentagon (which we shall call *P*) in the middle. This map can certainly be colored with four colors, but the point of Heawood's example is to show that Kempe's *method of proof* is incorrect.

Following Kempe, we now try to recolor two of the pentagon's neighbors in such a way that there is a spare color available for *P*. We notice first that the *blue* and *yellow* neighbors of *P* are connected by a *blue-yellow* chain of countries that separates the *red-green* part above *P* from the *red-green* part below *P*, as shown in figure (a) below. We can thus interchange the colors of the *red-green* part above *P* without affecting the *red-green* part below *P*, as in figure (b).

Alternatively, we could have carried out a different interchange of colors. The *blue* and *green* neighbors of *P* are connected by a *blue-green* chain of countries that separates the *red-yellow* part above *P* from the

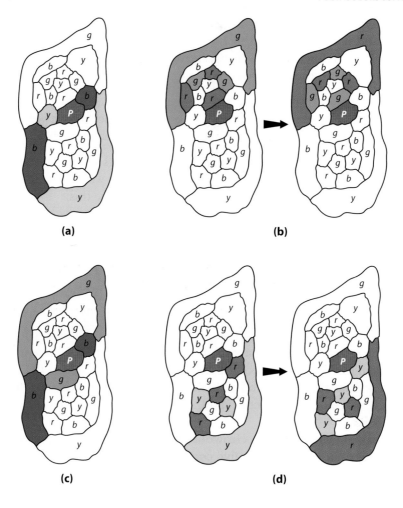

(a)   (b)

(c)   (d)

red-yellow part below *P*, as shown in figure (c) above. We can thus interchange the colors of the *red-yellow* part below *P* without affecting the
red-yellow part above *P*, as in figure (d).

Either of these color interchanges is permissible on its own, but
Kempe's error was in trying to do them simultaneously. For, if we interchange the colors in both the *red-green* part above *P* and the *red-yellow*
part below *P*, then the *green* country *A* and the *yellow* country *B* both
become *red*, which is not permissible.

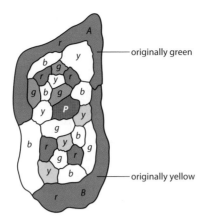

Thus Kempe's proof, which allowed the possibility of such simultaneous color interchanges, is fallacious. In Heawood's own words:

> But, unfortunately, it is conceivable that though either transposition would remove a red, both may not remove both reds. My map is an actual exemplification of this possibility, where either transposition prevents the other from being of any avail, by bringing the red and the other divisions into the same region; so that Mr. Kempe's proof does not hold, unless some modifications can be introduced into it to meet this case of failure.

The flaw in Kempe's argument proved to be a major one. Alfred Kempe admitted his error in the columns of the *Proceedings of the London Mathematical Society*, and on April 9, 1891, he discussed the difficulty at a meeting of the Society:

> My proof consisted of a method by which any map could be coloured with four colours. Mr. Heawood gives a case in which the method fails, and thus shows the proof to be erroneous. I have not succeeded in remedying the defect, though it can be shown that the map which Mr. Heawood gives can be coloured with four colours, and thus his criticism applied to my proof only and not to the theorem itself.

In fact, Heawood's map is not the simplest that he could have given, as was shown by two Belgian mathematicians. A more attractive example has the form of the buckyball C_{30} (see Chapter 3) and was described by Alfred Errera in 1921. A version of it appears below; as before, we can carry out either of the *red-green* and *red-yellow* color interchanges, but not both. In fact, Errera's map is much more than being a simpler example than Heawood's—it also fails Kempe's step-by-step method in the precise form in which he gave it.

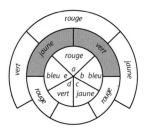

Errera's example   de la Vallée Poussin's example

Earlier, in 1896, Charles-Jean-Gustave-Nicolas de la Vallée Poussin, known particularly for his work on the distribution of prime numbers and his proof of the prime number theorem, had presented an example. He had read Edouard Lucas's translation of Kempe's proof in *Récréations Mathématiques* and, unaware of Heawood's paper, had located Kempe's error. Here, the pentagon appears as a single point and we can interchange *either* the countries colored *bleu* and *jaune* on the left *or* those colored *bleu* and *vert* on the right, but if we do both, then the two shaded countries both become *bleu*.

A SALVAGE OPERATION

Although Heawood was unable to mend the hole in Kempe's proof, he salvaged enough from Kempe's ideas to prove the *five-color theorem*. Although weaker than the four-color theorem, it is still a remarkable result:

FIVE-COLOR THEOREM

Every map can be colored with at most five colors in such a way that neighboring countries are colored differently.

To prove the five-color theorem, we imitate Cayley and Kempe's approach to the four-color problem in Chapters 4 and 5. That is, we begin by assuming that the five-color theorem is false, so that there are some maps that *cannot* be colored with five colors—and among these special maps that need five colors or more, we consider a minimal criminal—one with the smallest possible number of countries. This map *cannot* be colored with five colors, but any map with fewer countries *can* be colored with five colors.

We now use the "only five neighbors" theorem from Chapter 3. This theorem tells us that our map must contain a country with at most five neighbors—a digon, triangle, square, or pentagon. If there is a digon, triangle, or square, the proof is simple. We illustrate the method of proof for a map that contains a square. (The proofs for a digon or a triangle are similar but easier and are almost identical to the proof of the four-color theorem for a map containing a digon or a triangle given in Chapter 4.)

Suppose that our minimal criminal contains a square, as shown below. If we remove a boundary line from the square, merging the square with one of its former neighbors, we obtain a map with fewer countries. By our assumption, we can color this new map with five colors, *red, blue, green, yellow,* and *orange.*

We now reinstate the square. Since five colors are available and since the countries next to the square use only four of them, there must be a spare color for the square. Thus, we can color the minimal criminal with five colors, which contradicts our assumption. This shows that a minimal criminal cannot contain a square.

original map   obtain new map   color new map   color original map

If our minimal criminal contains a pentagon, then, as before, we remove a boundary line and merge the pentagon with its former neighbor, giving us a map having fewer countries. By our assumption, we can color this new map with five colors (below).

As before, we next reinstate the pentagon. But now the countries next to the pentagon may use all five colors, so there is no spare color to color the pentagon.

original map   obtain new map   color new map   try to color original map

To rescue the situation we use a Kempe-chain argument (see Chapter 5). We choose two of the surrounding colors that are not adjacent—say, *red* and *green*—and look at only those countries that are colored with these colors. (From here on, the proof is almost identical to the proof of the four-color theorem for a map containing a square.)

We look first at the *red* and *green* neighbors of the pentagon (which we call P). Each of these is the starting point for a part of the map consisting entirely of countries colored *red* or *green*. Now, are these two *red-green* parts separate from each other, or do they link up?

Two cases can arise:

case 1   case 2

CASE 1

In the first case, the *red* and *green* countries above P that can be reached from the *red* neighbor of P do not link up with the *red* and *green* countries below P that can be reached from the *green* neighbor of P. We can, therefore, interchange the colors of the *red* and *green* countries above P, as shown below, without affecting the coloring of the *red* and *green* countries below P. The pentagon P is then surrounded only by the colors *green, blue, yellow,* and *orange,* so P can be colored *red*. This completes the coloring of the map.

CASE 2

In this case, where the *red-green* part above P does link up with the *red-green* part below P, nothing is gained by interchanging the colors. So we turn our attention to the *blue* and *yellow* countries and to the *blue-yellow* parts of the map to the left and right of the pentagon P. Here, the *blue-yellow* part to the right of P is cut off from the *blue-yellow* part to the left of P, because the chain of *red* and *green* countries gets in the way, as illustrated below. We can, therefore, interchange the colors of the *blue* and *yellow* countries on the right of *P* without affecting the coloring of the *blue* and *yellow* countries on the left of *P*. The pentagon *P* is then surrounded only by the colors *yellow*, *red*, *green*, and *orange*, so *P* can be colored *blue*. This completes the coloring in the second case.

Thus, in either case, the minimal criminal can be colored with five colors, which contradicts our assumption. This shows that no minimal criminal can contain a pentagon and completes the proof of the five-color theorem.

COLORING EMPIRES

Percy Heawood was keen to extend the idea of map coloring beyond the original four-color problem. In the introduction to his paper in the *Quarterly Journal of Mathematics*, he pointed out that his main object was neither to call attention to the error in Kempe's proof nor to prove the five-color theorem: it was, rather, to contrast with the four-color problem "some remarkable generalizations of it, of which strangely the rigorous proof is much easier". One of these was the *empire problem*. It can be thought of as the problem of coloring a map with several "empires": each empire consists of a "mother country" and a number of "colonies" that must all be colored the same as the mother country.

In Chapter 1 we gave the following example of a map, due to Heawood, in which one country splits into two pieces and more than four colors are needed.

He now asked how many colors are needed if "any country may consist of two distinct portions but no more", so that each empire consists of a mother country and possibly a single colony. His simple example used five colors, but it is possible to construct maps that need many more colors than this. Using an argument based on Euler's formula, he showed that the number of colors required does not exceed twelve. But are there maps that actually need all twelve colors?

Heawood presented the example below, "obtained with much difficulty in a more or less empirical manner". It consists of twelve pairs of countries, with the property that each pair has a single boundary line in common with a country from every other pair. For example, of the two countries numbered *8*, one has boundary lines in common with countries numbered *1, 2, 6, 7, 9, 10*, and *12*, while the other has boundary lines in common with countries numbered with the remaining numbers, *3, 4, 5*, and *11*. Thus, there are twelve "mutually neighboring pairs of countries", which therefore need twelve colors.

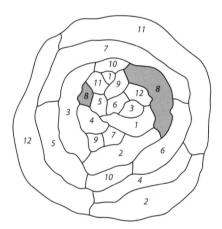

Again, using an argument based on Euler's formula, Heawood proved that if each empire consists of three distinct portions, then at most eighteen colors are needed; and, more generally, for any number *r* greater than 1, if each empire consists of *r* distinct portions, then at most 6*r* colors are needed. However, apart from the case *r* = 2 illustrated before, Heawood was unable to find maps for which these bounds are attained.

It was more than ninety years before any progress was made in solving the empire problem. In 1981, the case *r* = 3 was solved by Herbert Taylor, who constructed the eighteen-color map below. The solution for a general value of *r* was obtained three years later by Brad Jackson and Gerhard Ringel.

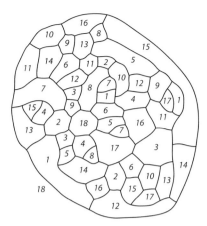

In 1959 Ringel posed a variation of the empire problem, known as the *earth-moon problem:* given two maps on the earth and the moon in which each earth-country has a moon-country as a colony, how many colors are needed to color their countries if each earth-country receives the same color as its moon-colony?

From our discussion of the empire problem with $r = 2$ we know that the required number of colors does not exceed twelve, and it has been proved that we require at least nine—but do there exist earth-moon maps for which ten, eleven, or twelve colors are needed? This question remains unanswered.

MAPS ON BAGELS

Heawood also investigated the coloring of maps drawn on surfaces other than the sphere. In his 1890 paper he presented an example of a map drawn on the surface of a torus (or bagel) in which each country adjoins every other country. Such a torus map, similar to Heawood's, is shown below—we came across it in Chapter 2 in connection with Möbius's problem of the princes.

contradicts our assumption. This shows that every map on a torus can be colored with seven colors.

We can summarize the preceding discussion thus:

> every map on a torus can be colored with seven colors
> —and there are torus maps that need seven colors.

Heawood proved these results in his paper and then tried to extend them to bagels with more than one hole (or "pretzels"). For example, how many colors are needed to color maps on the following two-holed torus?

By arguments similar to those used before, we can prove that the magic number is 8:

> every map on a two-holed torus can be colored with eight colors
> —and there are two-holed torus maps that need eight colors.

We can continue in this way:

> every map on a three-holed torus can be colored with nine colors
> —and there are three-holed torus maps that need nine colors;

> . . .

> every map on a ten-holed torus can be colored with fourteen colors
> —and there are ten-holed torus maps that need fourteen colors;

and so on.

For maps on a torus with an arbitrary number of holes, Heawood used Lhuilier's version of Euler's formula:

EULER'S FORMULA FOR THE h-HOLED TORUS

(number of countries) − (number of boundary lines)
+ (number of meeting points) = 2 − 2h

or, in symbols,

$$F - E + V = 2 - 2h.$$

This gives rise to a rather complicated-looking result:

every map on a torus with h holes can be colored with $H(h)$ colors,
where $H(h)$ is the number $[\tfrac{1}{2}(7 + \sqrt{1 + 48h})]$

Here, the square brackets mean that we round the number down if it is not already a whole number—for example,

$$[7] = 7 \text{ and } [9.99] = 9.$$

So, when $h = 1$ (a torus), the number of colors is

$$H(1) = [\tfrac{1}{2}(7 + \sqrt{49})] = [7] = 7;$$

when $h = 2$ (a two-holed torus), the number of colors is

$$H(2) = [\tfrac{1}{2}(7 + \sqrt{97})] = [8.42 \ldots] = 8;$$

and when $h = 10$ (a ten-holed torus), the number of colors is

$$H(10) = [\tfrac{1}{2}(7 + \sqrt{481})] = [14.46 \ldots] = 14.$$

These values agree with the numbers given before.

The number $H(h)$ is sometimes called the *Heawood number* of the h-holed torus. A table of values of $H(h)$ for small values of h is as follows:

number of holes, h	1	2	3	4	5	6	7	8	9	10
number of colors, $H(h)$	7	8	9	10	11	12	12	13	13	14

Unfortunately, even Heawood was capable of making mistakes. As we have seen, he correctly proved the formula for maps on the torus, showing that seven colors are sufficient, and he gave an example of a map on a torus that needs seven colors. For larger values of h, he indeed showed that every map on an h-holed torus can be colored with $[\tfrac{1}{2}(7 + \sqrt{1 + 48h})]$ colors. He then asserted that there are maps that re-quire this number of colors—but he neglected to prove this, comment-ing merely that

for highly connected surfaces it will be observed that there are generally con-tacts enough and to spare for the above number of divisions each to touch each.

Heawood's omission was a major blunder. It was first noticed in 1891, the year after Heawood published his paper, by Lothar Heffter of Giessen in Germany. Heffter managed to show that there are indeed maps that require $H(h) = \left[\frac{1}{2}(7 + \sqrt{1 + 48h})\right]$ colors when h = 2, 3, 4, 5, 6, and a few other values. But he was unable to prove the result in general. As we shall see in Chapter 9, seventy-seven years would pass before what came to be known as the *Heawood conjecture* was finally proved:

HEAWOOD CONJECTURE

For each positive number h, there is a map on the surface of an h-holed torus that requires $H(h) = \left[\frac{1}{2}(7 + \sqrt{1 + 48h})\right]$ colors.

Although neither Heawood nor Heffter lived to see the proof of the conjecture, their bagel-coloring activities seem to have bestowed longevity: Heawood died at the age of 94, while Heffter survived to the ripe old age of 99.

PICKING UP THE PIECES

Heawood's paper seems to have gone largely unnoticed. In the final volume of Edouard Lucas's *Récréations Mathématiques*, published posthumously in 1894, an extended version of Kempe's paper appeared, but with no mention of his error. In *L'Intermédiaire des Mathématiciens*, a newly formed Paris publication for the dissemination of mathematical problems, P. Mansion posed the four-color problem, seemingly unaware of any of the previous work on it. Replies were soon forthcoming—by H. Delannoy and A. S. Ramsey (citing papers by Kempe and Tait), by de la Vallée Poussin (presenting his simpler example of a map that exposes Kempe's error), and again by Delannoy (who did not understand de la Vallée Poussin's example and thought that Kempe had been right all along!).

The *Intermédiaire* discussion then moved on to Tait's assertion that all cubic polyhedra have Hamiltonian cycles—cycles passing through all the vertices. In 1898–99, the Danish mathematician Julius Petersen wrote two short notes, commenting on the connections between Tait's work and the four-color theorem and remarking that "M. Kempe only skimmed over the problem; he committed his error just where the difficulties began". Petersen concluded with a surprising opinion:

I know nothing with certainty, but if it came to a wager I would maintain that the theorem of the four colours is not correct.

Petersen is now mainly remembered for the *Petersen graph*, a diagram that appears below in its usual form, in Petersen's own 1898 drawing from the *Intermédiaire,* and in a version given twelve years earlier by Kempe. It can be shown that the Petersen graph has no Hamiltonian cycle and that it does not arise from a polyhedron.

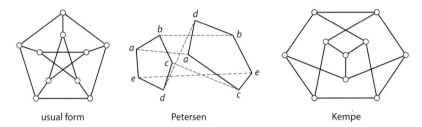

usual form   Petersen   Kempe

At no stage in the *Intermédiaire* correspondence were Heawood's achievements mentioned. Undeterred, Heawood continued to struggle with the four-color problem for the next sixty years. In 1898 he wrote a paper that included a particularly useful development of Tait's connection (see Chapter 6) between coloring the boundary lines of a cubic map with three colors and coloring its countries with four colors:

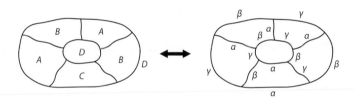

Heawood now focused on the meeting points, to which he assigned the value 1 if the colors α, β, and γ occur in clockwise order, and -1 if they occur in counterclockwise order. For example, the top point in the diagram above is assigned the value 1, because the colors α, β, and γ appear there in clockwise order. This gives the following arrangement of 1s and -1s.

Heawood showed that the sum of the numbers around each country of the map must always be divisible by 3: here, the sum is 3 for the country in the middle, 0 for each country in the ring, and −3 for the exterior country. Turning the argument around, he then showed that

> if we can assign the numbers 1 and −1 to the meeting points of a cubic map so that the sum of the numbers around each country is divisible by 3, then we can colour the boundary lines with three colours, and the countries with four colours.

Thus, solving the four-color problem amounts to showing that we can always assign the numbers 1 and −1 to the meeting points in this way. Heawood was so fascinated by this approach that he wrote five further papers on it, but without achieving the ultimate success that he so much craved.

There is a simple consequence of Heawood's result, which Heawood himself seems to have been the first to notice. As he observed in his 1898 paper: "If the number of neighbours of each country is divisible by 3, then the map can be coloured with four colours." To see why, assign the number 1 to every meeting point. Then, by the condition on the neighbors, the sum of the numbers around each country must be divisible by 3, and so the colors α, β, and γ must appear in clockwise order around each point. We can then obtain the required coloring of the countries, as explained in Chapter 6.

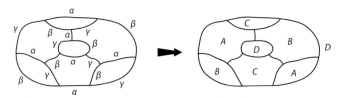

We conclude this chapter with a further result of Heawood's. In a paper of 1936, written when he was in his mid-70s, he attempted to estimate the probability that the four-color problem was true—or, more precisely, the probability that a random map with n countries could be colored with four colors. His arguments were somewhat crude, but they indicated that the probability of failure did not exceed about $e^{-4n/3}$, where e is the exponential number 2.71828 . . . and n is fairly large. By 1936 the four-color theorem was known to be true for all maps with $n = 27$ countries, as we shall see in Chapter 8, so his arguments would imply that the likelihood of failure was less than one in a million billion. It followed that if the four-color theorem were false, then a minimal criminal would be very hard to find.

8

Crossing the Atlantic

The end of the nineteenth century marks a watershed in the fortunes of the four-color problem. Kempe's solution had been shown to be faulty, and nothing had turned up to take its place. It was time for some new ideas to burst upon the scene.

The feeling was also beginning to emerge in some quarters that a solution to the four-color problem had not been forthcoming because no really good mathematicians had worked on it. Indeed, a story is told about the distinguished German number theorist Hermann Minkowski in the first decade of the twentieth century. While lecturing on topology at Göttingen University, he mentioned the four-color problem:

'This theorem has not yet been proved, but that is because only mathematicians of the third rank have occupied themselves with it', Minkowski announced to the class in a rare burst of arrogance. 'I believe I can prove it.'

He began to work out his demonstration on the spot. By the end of the hour he had not finished. The project was carried over to the next meeting of the class. Several weeks passed in this way. Finally, one rainy morning, Minkowski entered the lecture hall, followed by a crash of thunder. At the rostrum, he turned towards the class, a deeply serious expression on his face.

'Heaven is angered by my arrogance', he announced. 'My proof of the Four Colour Theorem is also defective.' He then took up the lecture on topology at the point where he had dropped it several weeks before.

With the turn of the century, things began to change. The four-color saga ceased to be a largely British story when several American mathematicians—George Birkhoff, Oswald Veblen, Philip Franklin, Hassler Whitney, and others—contributed to a major new chapter. Through their work, two ideas gradually emerged that were to prove decisive. Both ideas had been implicit in Kempe's paper—the concepts of an *unavoidable set* and a *reducible configuration*.

TWO FUNDAMENTAL IDEAS

In Chapter 3 we proved the "only five neighbors" theorem—that every map has at least one country with five or fewer neighbors. It follows, in particular, that every cubic map must contain at least one of the following:

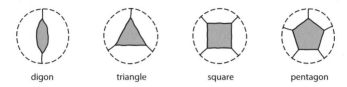

digon   triangle   square   pentagon

So whenever we draw a cubic map, we have to use at least one country from this collection. We describe such a collection of countries as an *unavoidable set* because we cannot avoid it—in any cubic map at least one of them must appear somewhere. As we shall see later, another unavoidable set is this one:

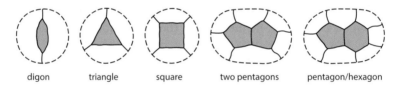

digon   triangle   square   two pentagons   pentagon/hexagon

Thus, if a cubic map contains no digon, triangle, or square, then not only must it contain a pentagon, it must contain either two adjacent pentagons or a pentagon adjacent to a hexagon. So why should we be interested in unavoidable sets?

In trying to prove the four-color theorem, our approach has been to consider a minimal criminal—a cubic map with the smallest possible number of countries that *cannot* be colored with four colors—and to prove that minimal criminals cannot exist. In Chapter 4 we showed that no minimal criminal can contain a digon or triangle, since if it did, any coloring of the rest of the map with four colors could immediately be extended to include the digon or triangle as well. In Chapter 5 we gave Kempe's proof that no minimal criminal can contain a square—if there were such a country, then, by using a Kempe-chain argument that interchanges two of the colors in part of the map, we could free up a color for the square. But Kempe was unable to prove satisfactorily that no minimal criminal can contain a pentagon: the error in his proof was pointed out by Heawood, as we saw in Chapter 7.

So the smallest country there can be in a minimal criminal is a pentagon. Moreover, as we saw in our discussion of the counting formula in Chapter 3, the map must contain at least twelve pentagons. It follows that a minimal criminal must have at least twelve countries. However, if a cubic map has *exactly* twelve countries, then they must all be pentagons, and we obtain the map of the dodecahedron. But this map *can* be colored with four colors, as shown below, so it cannot be a minimal criminal. It follows that every map with up to twelve countries can be colored with four colors, and hence that a minimal criminal must have at least thirteen countries.

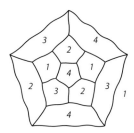

A *reducible configuration* is any arrangement of countries that cannot occur in a minimal criminal—so a digon, a triangle, and a square are all reducible configurations. If a map contains a reducible configuration, then any coloring of the rest of the map with four colors can be extended, after any necessary recoloring, to a coloring of the entire map. If Kempe had also been able to prove that the pentagon is reducible, then the four-color problem would have been solved.

The rest of this book is concerned with the attempt to find *an unavoidable set of reducible configurations*. Finding such a set proves the four-color theorem: since the set is unavoidable, every map must contain at least one of the configurations, but each configuration is reducible and so cannot be contained in a minimal criminal. Thus, no minimal criminal can exist. But how do we find unavoidable sets, and how do we find reducible configurations?

FINDING UNAVOIDABLE SETS

We have just seen how Kempe failed to deal with the pentagon. Can we replace the pentagon by some other arrangements of countries that we can handle more successfully?

The first attempts to do this were made by Paul Wernicke, an American mathematician who received a doctoral degree from Göttingen

University and then recrossed the Atlantic to take up a professorship at the University of Kentucky. At one stage he was in charge of the university's military training, holding a commission as a colonel in the Kentucky militia; it is not recorded whether, in common with another Kentucky colonel, he had a liking for fried chicken.

In August 1897 Wernicke presented a paper entitled "On the solution of the map-color problem" at the Fourth Summer Meeting of the American Mathematical Society, held in Canada at the University of Toronto. The Society's *Bulletin* included a short abstract of his paper, from which it appears that he was concerned with developing Tait's connection (described in Chapter 6) between coloring the boundary lines of a cubic map with three colors and coloring its countries with four colors. Wernicke's idea seems to have been to add new countries to a map so as to turn it into one he could manage:

> Given a map correctly colored and with its frontiers marked, the author proves that any triangles, quadrangles, and pentagons can be introduced and correctly marked at the same time. The main theorem then follows by induction.

However, it is unlikely that Wernicke's proof of the four-color theorem, by the method of induction, was any more successful than Tait's had been seventeen years earlier.

Wernicke's next attempts were much more productive. In a lengthy paper written in Göttingen in May 1903, he proved that a cubic map that contains no digon, triangle, or square must contain either two adjacent pentagons or a pentagon adjacent to a hexagon. Thus, as we saw before, the following configurations of countries form an unavoidable set:

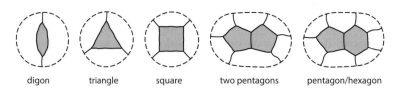

digon   triangle   square   two pentagons   pentagon/hexagon

To explain why Wernicke's result is true, we use a modern approach based on the *method of discharging*. This method was originated by Heinrich Heesch (pronounced "haish") and first published in 1969; Heesch's work takes center stage in Chapter 9. The term *discharging* was introduced by Wolfgang Haken, whose contributions to the eventual solution of the four-color problem are described in Chapter 10.

We illustrate the method of discharging by showing that the preceding set of configurations—the digon, triangle, square, pair of adjacent pentagons, and pentagon adjacent to a hexagon—forms an unavoidable set. To do this, we first assume that we have a cubic map that contains none of them and then seek a contradiction. The assumption tells us that no pentagon can be adjacent to a digon, a triangle, or a square (because there are none) or to another pentagon or to a hexagon. Thus, each pentagon can adjoin only countries bounded by at least seven edges.

We now assign to each country a number—which we can think of as an electrical charge. This we do as follows: to each country with k boundary lines, we assign a charge of $6 - k$, so that

each pentagon ($k = 5$) receives a charge of 1;
each hexagon ($k = 6$) receives zero charge;
each heptagon ($k = 7$) receives a charge of -1;
each octagon ($k = 8$) receives a charge of -2;

and so on.

So if the map has C_5 pentagons, C_6 hexagons, C_7 heptagons, and so on, then the total charge on the map is

$$(1 \times C_5) + (0 \times C_6) + (-1 \times C_7) + (-2 \times C_8) + (-3 \times C_9) + \ldots$$
$$= C_5 - C_7 - 2C_8 - 3C_9 - \ldots \qquad (1)$$

We now recall the counting formula from Chapter 3: if a cubic map has C_2 digons, C_3 triangles, C_4 squares, and so on, then

$$4C_2 + 3C_3 + 2C_4 + C_5 - C_7 - 2C_8 - 3C_9 - \ldots = 12.$$

Since our map has no digons, triangles, or squares, we have $C_2 = C_3 = C_4 = 0$, and the counting formula simplifies to

$$C_5 - C_7 - 2C_8 - 3C_9 - \ldots = 12. \qquad (2)$$

Comparing equations (1) and (2), we find that the total charge on the map is 12, a positive number.

We now move the charges around the map in such a way that no charge is created or destroyed—this is called *discharging the map* and is akin to the conservation of electrical charge. One way of doing this is to transfer one-fifth of a unit of charge from each pentagon to each of its five negatively charged neighbors—those with seven or more sides. The diagram below shows an example of this:

The result of the discharging is that the total charge on the map remains 12, but each pentagon now has zero charge, and each hexagon still has zero charge.

What happens to a heptagon? For a heptagon with initial charge -1 to receive enough contributions of $\frac{1}{5}$ to acquire positive charge, it would need at least six neighboring pentagons, as shown below, supplying $\frac{6}{5}$ in total—but then at least two of these pentagons would have to be adjacent, which is not allowed. After the discharging, then, each heptagon retains its negative charge.

And what happens to an octagon? For an octagon with initial charge -2 to receive enough contributions of $\frac{1}{5}$ to acquire positive charge, it would need at least eleven neighboring pentagons, which is clearly impossible. So, after the discharging, each octagon retains its negative charge—and so does each nonagon, decagon, and so on.

Thus, after the discharging, each country of our map has zero or negative charge. But this is inconsistent with the map having a total charge of 12. This contradiction proves that every cubic map must contain at least one of the above configurations of countries—a digon, a triangle, a square, two adjacent pentagons, and a pentagon adjoining a hexagon—and so these five configurations form an unavoidable set.

The idea of an unavoidable set was later developed by two other mathematicians. The first was Philip Franklin, who wrote a doctoral thesis on map coloring at Princeton University and who later became a distinguished mathematician at the Massachusetts Institute of Technology (MIT). He was a brother-in-law of Norbert Wiener, the inventor of cybernetics—the comparative study of communication processes and automatic control systems.

In 1920 Franklin presented part of his thesis to the National Academy of Sciences. His presentation included the result, which he derived from the counting formula, that every cubic map must contain at least one digon, triangle, square, or one of the following:

a pentagon adjacent to two other pentagons;
a pentagon adjacent to a pentagon and a hexagon;
a pentagon adjacent to two hexagons.

This gives rise to the unavoidable set with nine configurations shown below.

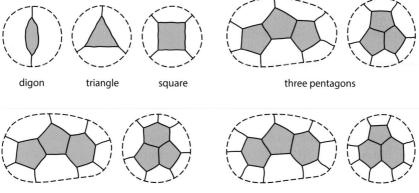

digon   triangle   square   three pentagons

two pentagons and a hexagon   a pentagon and two hexagons

The other provider of unavoidable sets was the French mathematician Henri Lebesgue, better known for developing the *Lebesgue integral* used in mathematical analysis. In 1940, the year before his death, Lebesgue wrote a paper on some simple consequences of Euler's formula in which he used the counting formula to construct a number of interesting new unavoidable sets.

By modifying the method of discharging, we can show that many sets of configurations are unavoidable, but the details of the discharging process may vary from one situation to another. Earlier we trans-

ferred one-fifth of a unit of charge from each pentagon to each negatively charged neighbor, but in other cases it may be more advantageous
to transfer one-fourth or one-third of a unit or to distribute the charge
on each pentagon equally among *all* its neighbors.

As the twentieth century progressed, unavoidable sets with thousands of configurations were constructed. To deal with such enormous
sets it proved necessary to keep on modifying the discharging process
until it could deal with every possible case that arose. In Chapter 10 we
shall see how this was done.

FINDING REDUCIBLE CONFIGURATIONS

We have seen before that the digon, triangle, and square are all examples of reducible configurations—if they appear in a map, then any coloring of the rest of the map with four colors can be extended, possibly
after some recoloring, to include them. But so far our supply of reducible configurations has been very restricted. This situation changed dra-
matically in 1913 with the publication of a paper by George David
Birkhoff.

Birkhoff, one of the outstanding American scholars of the early twentieth century, made substantial contributions to many areas of mathematics. He studied at the University of Chicago and at Harvard University, later obtaining positions at the University of Wisconsin and at
Princeton University before returning to Harvard in 1912 to spend the
rest of his productive life.

While at Princeton, Birkhoff attended seminars by the celebrated geometer Oswald Veblen, who took a keen interest in the four-color problem. On April 27, 1912, Veblen read a paper to the American Mathematical Society, extending the ideas of Percy Heawood's 1898 paper and
placing them in the context of a particular type of geometry in which
each line contains exactly four points. Veblen was later to supervise
Philip Franklin's doctoral thesis on the coloring of maps.

From this time onward, Birkhoff regarded solving the four-color problem as one of his greatest aspirations, even though he was later to regret
the amount of time he had spent on it. In 1913 he published his groundbreaking paper in the *American Journal of Mathematics*, the journal in
which Kempe's famous solution had appeared more than thirty years before. This paper, "The reducibility of maps", written while Birkhoff was
still at Princeton, gave a systematic treatment of Kempe-chain arguments
and laid the groundwork for all later developments in this direction.

George David Birkhoff (1884–1944).

Birkhoff considered *rings of countries* in minimal criminals. To see what is involved, suppose that a map contains a ring of three countries, with at least one other country inside the ring and at least one other country outside it (the ring of countries is colored orange).

Since the entire map is a minimal criminal, we can color the ring and its interior with four colors, and we can also color the ring and its exterior. We can then match up the three colors on the ring (by renaming the colors, if necessary) and thereby obtain a coloring of the whole map (see overleaf). It follows that rings of three countries are reducible and so cannot appear in a minimal criminal.

Notice that this method generalizes Kempe's ideas. Instead of removing just one country and coloring the rest of the map with four colors, we now remove several countries at a time—in this case, the countries

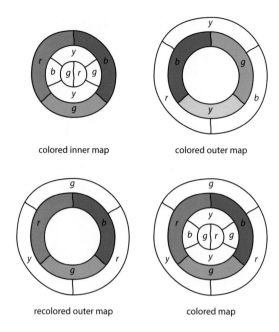

colored inner map   colored outer map

recolored outer map   colored map

inside or outside the ring. We then color the rest of the map with four colors and match up the separate colorings on the ring to give a coloring for the entire map.

The corresponding problem for rings of four countries is a little more difficult because such a ring can be colored with two, three, or four colors, and it may not be easy to match up the colorings. The following diagrams show ring colorings with two and three colors that cannot be matched up directly:

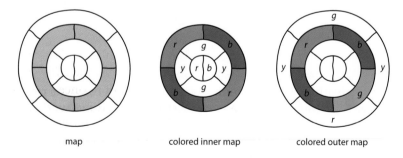

map   colored inner map   colored outer map

However, by switching the colors in appropriate two-colored Kempe chains, Birkhoff showed that these difficulties can always be overcome. He deduced that rings of four countries are reducible and so cannot appear in a minimal criminal.

Birkhoff then extended his arguments to rings of five countries and succeeded in every case except one: a single pentagon surrounded by a ring of five countries—the case that had defeated Kempe. Birkhoff's arguments also extended to certain rings of six countries, but these proved to be much more problematic. While carefully analyzing them, Birkhoff became increasingly uncertain of the status of the four-color problem, and in his paper he concluded that any of the following possibilities seemed plausible:

1. There exist maps that cannot be colored in four colors, a leading feature of the simplest one of them perhaps being a ring of six regions with more than three regions on each side. By the method of reduction, one will always be led either to a coloring of the given map or to one or more maps that cannot be colored.
2. All maps can be colored in four colors and a set of reducible rings can be found, one of which exists in every map.
3. All maps can be colored in four colors but only by means of reductions of a more extensive character, applicable to sets of regions bounded by any number of rings.

More than thirty years later, in an important but highly technical paper, Arthur Bernhart of the University of Oklahoma managed to complete Birkhoff's work on rings of six countries. The story is told that shortly after Bernhart married, his new wife encountered Mrs. Birkhoff at a mathematics meeting. Mrs. Birkhoff quizzed Mrs. Bernhart:

Tell me, did *your* husband make you draw maps for him to color on your honeymoon, as mine did?

Whatever the answer, four colors clearly ran in the blood, as the Bernharts' son Frank also became a well-known writer on the four-color problem.

In the next section we shall outline Birkhoff's argument for showing that a particular six-ring configuration of countries—the *Birkhoff diamond*—is reducible. This is an important configuration, once described as enjoying "as much renown in graph theory as the Kohinoor diamond does in fictional criminal mysteries". After this the floodgates were open, and mathematicians on both sides of the Atlantic began to develop Birkhoff's ideas, generating reducible configurations galore.

The four-color problem also became a fashionable topic for research degrees, and a number of students obtained doctoral degrees for dissertations on map coloring. One of these students, whom we mentioned earlier, was Philip Franklin. In his Princeton thesis "On the map color problem" he showed that each of the following configurations is reducible and so cannot appear in a minimal criminal:

a pentagon in contact with three pentagons and one hexagon;
a pentagon surrounded by two pentagons and three hexagons;
a hexagon surrounded by four pentagons and two hexagons;
and any n-sided polygon in contact with $n - 1$ pentagons.

By applying the counting formula, he was then able to deduce that every map with up to twenty-five countries can be colored with four colors and, therefore, that a minimal criminal must have at least twenty-six countries.

Another young researcher was Alfred Errera, who wrote his dissertation "Du coloriage des cartes" at the University of Brussels. Errera extended Franklin's results, proving in particular that a minimal criminal must contain at least thirteen pentagons and that it cannot contain only pentagons and hexagons.

Later mathematicians continued this work, obtaining further reducible configurations and using the counting formula to prove the four-color theorem for maps with more and more countries. In particular, Clarence Reynolds of West Virginia showed in 1926 that four colors are sufficient for all maps with up to twenty-seven countries. Franklin then increased this number to thirty-one in 1938, and two years later C. E. Winn, of the Egyptian University in Cairo, increased it to thirty-five, where it stuck for a quarter of a century. Thus, a hundred years after De Morgan's original letter of 1852, it had been proved that all maps with up to thirty-five countries can be colored with four colors—but there was still a long way to go.

We conclude this general discussion of reducible configurations by remarking that several of the preceding ideas had surfaced much earlier, in a most unlikely place. Like so many others, the French writer and poet Paul Valéry was fascinated by the four-color problem, and his diaries for 1902 were later found to contain a dozen pages of substantial work on configurations of countries that were later to be explored by Birkhoff, Franklin, and Winn.

COLORING DIAMONDS

In this section we illustrate Birkhoff's methods by proving that the so-called Birkhoff diamond of four pentagons is reducible and so cannot appear in a minimal criminal.

Birkhoff diamond

Suppose that we have a minimal criminal that contains the Birkhoff diamond. Removing the diamond yields a new map with fewer countries. We assume that this new map can be colored with four colors and try to extend the coloring to the pentagons in the diamond.

If the countries in the ring surrounding the diamond are numbered 1, 2, 3, 4, 5, and 6, as shown, it turns out that there are essentially thirty-one different ways in which they can be colored with the four colors *red*, *green*, *blue*, and *yellow*. These colorings are as follows (the reason for the asterisks is given below):

rgrgrg	*rgrbrg**	*rgrbgy**	*rgbrgy*	*rgbryb*	*rgbgbg**	*rgbyrg*	*rgbygy**
*rgrgrb**	*rgrbrb*	*rgrbyg**	*rgbrbg**	*rgbgrg**	*rgbgby*	*rgbyrb*	*rgbybg**
rgrgbg	*rgrbry*	*rgrbyb**	*rgbrby*	*rgbgrb**	*rgbgyg*	*rgbyry**	*rgbyby**
*rgrgby**	*rgrbgb**	*rgbrgb*	*rgbryg*	*rgbgry**	*rgbgyb*	*rgbygb*	

Notice that we cannot include colorings such as *rgygbr*, in which the two countries colored *red* appear together, and we have also omitted colorings such as *rgrgry*, since this is essentially the same coloring as *rgrgrb* (on recoloring the final *yellow* as *blue*).

Consider the coloring *rgrgrb*. This can be extended directly to the diamond, as shown below, and for this reason it is called a *good coloring*. In the same way, all the preceding asterisked colorings are good colorings: try some of them and see!

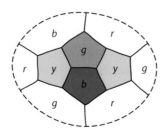

The coloring *rgrbrb* is not a good coloring, but by using Kempe-chain changes of color that interchange the colors *red* and *yellow* or *green* and *blue*, it can be converted into one of the good colorings *rgrgrb*, *rgrbrg*, or *rgrbyb*. For example, if there is a *red-yellow* chain connecting countries 3 and 5, then we can interchange the colors in the *blue-green* chain containing country 4 in order to recolor country 4 *green*. Similarly, if there is a *red-yellow* chain connecting countries 1 and 5, then we can interchange the colors in the *blue-green* chain containing country 6 in order to recolor country 6 *green*. However, if there is no *red-yellow* chain connecting countries 3 and 5 or 1 and 5, then we can interchange the colors in the *red-yellow* chain containing country 5 in order to recolor country 5 *yellow*. (These three situations are illustrated below.) Thus, the coloring *rgrbrb* can be converted into a good coloring.

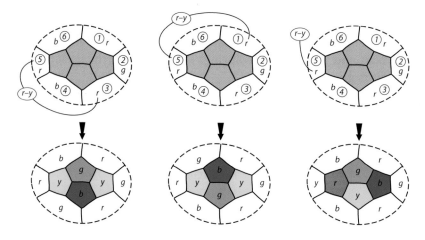

The coloring *rgrbry* is also not a good coloring, but by using Kempe-chain changes that interchange the colors *red* and *green* or *blue* and *yellow*, it can be converted into either *rgrbgy* (which is good) or into *rgrbrb* (which can be made good, as we have just seen). This is because, if there is a *blue-yellow* chain connecting countries 4 and 6, then we can interchange the colors in the *red-green* chain containing country 5 in order to recolor country 5 *green*. However, if there is no *blue-yellow* chain connecting countries 4 and 6, we can interchange the colors in the *blue-yellow* chain containing country 6 in order to recolor country 6 *blue*. (These two situations are illustrated below.) Thus, the coloring *rgrbry* can also be converted into a good coloring.

It turns out that each of the thirty-one possible colorings of the ring is either a good coloring or can be converted into a good coloring by

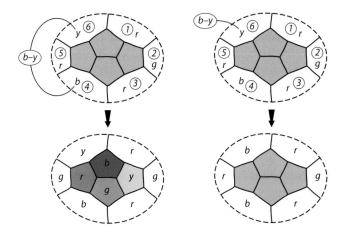

suitable Kempe-chain changes of color. Thus, all thirty-one colorings of the ring can be extended to the diamond, so the Birkhoff diamond is reducible.

In fact, we do not have to consider all thirty-one colorings. If we modify the map by removing five of the boundary lines, as illustrated below, then we obtain a new map with fewer countries, which can, therefore, be colored with four colors:

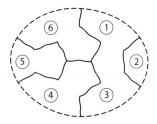

This modification corresponds to eliminating all those colorings in which countries 1 and 3 have different colors or countries 4 and 6 have the same color. The effect of this is to eliminate all the thirty-one colorings except *rgrgrb*, *rgrgby*, *rgrbrg*, *rgrbgy*, *rgrbyg*, and *rgrbry*. The first five of these are good colorings, and the last can be converted into a good coloring, as we saw before. Thus, we can color the new map with four colors, so the configuration is reducible.

Heinrich Heesch introduced the term *D-reducible configuration* for configurations of countries for which every coloring of the immediately surrounding ring of countries is a good coloring, or can be converted into a good coloring by a succession of Kempe-chain changes of color. Thus, a digon, a triangle, and a square are all *D*-reducible configurations, as shown by Kempe, and so is the Birkhoff diamond, as we have just seen. Heesch also used the term *C-reducible configuration* for configurations that can be proved reducible after they have been modified in some way, as illustrated before. The concepts of *D*-reducible and *C*-reducible configurations will reappear in Chapters 9 and 10.

HOW MANY WAYS?

Birkhoff once remarked that almost every great mathematician had worked on the four-color problem at one time or another, and in one of his last papers, a 97-page study of map coloring written with D. C. Lewis and published posthumously, he classified all previous investigations into two types: *qualitative approaches* and *quantitative approaches*.

In the qualitative approach, the aim is to show that all maps of a certain type can be colored with four colors. Here, Kempe chains play an important role, and the results for reducible configurations obtained by Birkhoff, Franklin, Errera, Reynolds, and Winn can be regarded among the successes of this approach.

In the quantitative approach we can use any number of colors, and we want to find the number of ways in which a given map can be colored with these colors. Our object is to show that if four colors are available, then the number of ways in which the map can be colored is a positive number. The quantitative approach was introduced by Birkhoff while he was still at Princeton University, and his seminal paper was published in the *Annals of Mathematics* in 1912, the year before his paper on reducible configurations and the Birkhoff diamond.

To see what is involved, we start with this simple map:

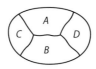

Birkhoff took the number of colors to be λ (lambda), where λ is any number larger than 2. Then country A can be colored with any of these λ colors. Since country B is a neighbor of country A, it can then be colored with any of the remaining $\lambda - 1$ colors. Finally, since countries C and D are both neighbors of countries A and B but not of each other, they can each be colored with any of the remaining $\lambda - 2$ colors. Thus, the total number of ways of coloring all the countries of this map is

$$\lambda \times (\lambda - 1) \times (\lambda - 2)^2.$$

For example, if we have four colors ($\lambda = 4$), then the number of ways of coloring this map is $4 \times 3 \times 2^2 = 48$, and if we have ten colors ($\lambda = 10$), then the number of ways of coloring it is $10 \times 9 \times 8^2 = 5760$.

Birkhoff used the symbol $P(\lambda)$ to denote the number of ways of coloring a map with λ colors; thus, for the preceding map,

$$P(\lambda) = \lambda \times (\lambda - 1) \times (\lambda - 2)^2,$$

which we can multiply out to give

$$P(\lambda) = \lambda^4 - 5\lambda^3 + 8\lambda^2 - 4\lambda.$$

As a check we can substitute $\lambda = 4$ into this formula, giving

$$P(4) = 4^4 - (5 \times 4^3) + (8 \times 4^2) - (4 \times 4)$$

$$= 256 - (5 \times 64) + (8 \times 16) - 16 = 48,$$

as before.

The preceding expression $P(\lambda)$, involving multiples of powers of λ, is called a *polynomial in λ*, and the numbers 1, −5, 8, and −4 preceding these powers are its *coefficients*. Birkhoff proved that the number of ways of coloring any map with λ colors is always a polynomial in λ, which he called the *chromatic polynomial* of the map, and he derived expressions for its coefficients in terms of properties of the map. Notice that if $P(4)$ is a positive number, then the map can be colored with four colors.

Another result that Birkhoff proved was pointed out to him by his Harvard University doctoral student Hassler Whitney. If we look at the coefficients 1, −5, 8, and −4 of this chromatic polynomial, we notice that these coefficients alternate in sign between positive and negative. It turns out that this is always the case:

for any map, the coefficients of the chromatic polynomial alternate in sign.

In a paper presented to the American Mathematical Society on October 25, 1930, Whitney proved this result and then related his conclusions to properties of the map.

Birkhoff always hoped that he could solve the four-color problem by investigating the properties of these chromatic polynomials, $P(\lambda)$. He wrote four papers about them, including the lengthy and very technical paper with Lewis mentioned earlier. Among the results that he obtained was the inequality

$$P(\lambda) \geq \lambda \times (\lambda - 1) \times (\lambda - 2) \times (\lambda - 3)^{n-3},$$

which is true for any map with n countries and where λ is any positive integer *except* 4. If he could also have proved this result for $\lambda = 4$, then he would have shown that

$$P(4) \geq 4 \times 3 \times 2 \times 1^{n-3} = 24,$$

which is a positive number. He would therefore have proved that every map can be colored with four colors (in at least twenty-four ways, in fact), thereby solving the four-color problem.

Since Birkhoff's death there has been much further work on chromatic polynomials. They have been calculated for large numbers of maps, and their properties have been investigated in great detail. We conclude this chapter with a bizarre result due to Bill Tutte and his colleagues, derived in the late 1960s.

If a map is large, then we almost certainly need four colors to color its countries. This means that it cannot be colored with one, two, or three colors, and so $P(1)$, $P(2)$, and $P(3)$ are all 0, whereas $P(4)$ is greater than 0. Now, are there any numbers x, other than 1, 2, and 3, for which $P(x) = 0$? If there are, then investigating them may give us an insight into the behavior of chromatic polynomials in general.

Tutte's result is related to the *golden ratio*, given by $\frac{1}{2}(1 + \sqrt{5})$ = 1.618034 . . . This number, sometimes denoted by the Greek letter τ (tau), appears throughout mathematics. It is the ratio of a diagonal and a side of a regular pentagon, and a "golden rectangle" with sides in this ratio is sometimes considered to have the most pleasing shape—neither too thin nor too fat:

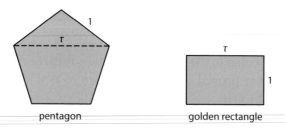

pentagon   golden rectangle

The golden ratio τ has interesting numerical properties. For example, if we calculate its reciprocal and its square, we get exactly the same pattern of decimals:

$$\tau = 1.618034 \ldots, \quad \tfrac{1}{\tau} = 0.618034 \ldots, \text{ and } \tau^2 = 2.618034 \ldots$$

These results can easily be deduced from the quadratic equation $x^2 = x + 1$, whose solutions are τ and $-1/\tau$.

In 1969, Gerald Berman and Bill Tutte noticed that for large cubic maps, the value of the chromatic polynomial $P(x)$ tends to lie extremely close to 0 when $x = \tau^2$, usually with an agreement to several decimal places. In the following year, Tutte supported these observations theoretically by proving that, when $x = \tau^2$, the value of $P(x)$ can never exceed τ^{5-n}, where n is the number of countries in the map. For example, for cubic maps with twenty countries, the value of $P(\tau^2)$ cannot exceed $\tau^{5-20} = \tau^{-15}$, which is about 0.0007, while for cubic maps with thirty countries, the value of $P(\tau^2)$ cannot exceed $\tau^{5-30} = \tau^{-25}$, which is about 0.000006. Forty years later, it is still not clear what implications these results have for the four-color problem.

9

A New Dawn Breaks

By the middle of the twentieth century, much progress had been made on solving the four-color problem. Only three papers had appeared on unavoidable sets of configurations, by Wernicke (1904), Franklin (1922), and Lebesgue (1940), but dozens of reducible configurations were being discovered, following on from the work of Birkhoff. Also, as we saw in Chapter 8, Franklin, Winn, and others had demonstrated that if the four-color theorem were false, then any minimal criminal would be complicated, containing more than thirty-five countries.

Meanwhile, large numbers of papers were appearing in the closely related area of graph theory—the study of "linkages", mentioned by Kempe and Tait and later developed by Hassler Whitney and others. Graph theory was proving to be of great importance through its applications, particularly in a wide range of network problems on which great advances had been made during the 1950s. Many mathematicians were being attracted to the subject as an area of interest in its own right, largely through the appearance of important textbooks. These included the first major book on the subject, Dénes König's *Theorie der endlichen und unendlichen Graphen* (*Theory of Finite and Infinite Graphs*), which appeared in 1936, and later textbooks on graph theory by Claude Berge in France and by Oystein Ore, Robert Busacker and Thomas Saaty, and Frank Harary in the United States.

The 1960s also proved to be an exciting time for map coloring. In 1967 the first major book devoted exclusively to map coloring, Oystein Ore's authoritative and influential *The Four-Color Problem*, was published. The next year, Ore and his research student, Joel Stemple, extended the earlier methods of Franklin and Winn to prove that all maps with up to forty countries can be colored with four colors. They needed to consider so many special cases in their proof that the full details could not be published but had to be deposited with the Mathematics Department Library at Yale University. (When the general proof finally appeared, it was also criticized for not being publishable in full.)

BAGELS AND TRAFFIC COPS

One undoubted success story around this time was the proof in 1968 of the Heawood conjecture by Gerhard Ringel from Germany and Ted Youngs of California. As we saw in Chapter 7, Percy Heawood had proved that every map drawn on the surface of an h-holed torus can be colored with $H(h) = [\frac{1}{2}(7 + \sqrt{1 + 48h})]$ colors, but he failed to prove that there are maps on an h-holed torus that actually require this number of colors, when h is larger than 1. As Gerhard Ringel later wryly commented:

> In 1890 P. J. Heawood published a formula which he called the Map Color Theorem. But he forgot to prove it. Therefore the world of mathematicians called it the Heawood Conjecture. In 1968 the formula was proven and therefore again called the Map Color Theorem.

The proof was a veritable tour de force. Just as Möbius's problem of the five princes is equivalent to the problem of connecting five palaces with noncrossing roads (see Chapter 2), so Heffter in 1891 showed that the Heawood conjecture is equivalent to a problem about the connecting of n points with noncrossing lines on a torus with a certain number of holes. This certain number involved the fraction $\frac{1}{12}$, and the denominator 12 turned out to be highly significant: in fact, the eventual proof of the Heawood conjecture split into twelve completely separate cases, depending on the remainder when n is divided by 12.

Ted Youngs (left) and Gerhard Ringel in 1968.

By the summer of 1967, all but three of these cases had been settled, and Ted Youngs invited Ringel to visit California for the academic year 1967–68 to work on them. They struggled for several months, and eventually the proof was completed. Great rejoicing!

Solving map-coloring problems can sometimes yield unexpected benefits. Shortly after the proof of the Heawood conjecture had been announced, Mrs. Ringel was driving along the California expressway and was stopped by a traffic cop for a minor traffic violation. As soon as he found that the culprit's name was Ringel, the cop asked, "Is your husband the one that solved the Heawood conjecture?" Mrs Ringel, surprised, admitted that he was and was duly let off with only a warning. As it happened, the cop's son had been in Ted Youngs's calculus class when the proof of the Heawood conjecture had been announced.

Before leaving the Heawood conjecture, notice that if we substitute $h = 0$ into Heawood's formula, then we obtain the correct answer for the number of colors required for maps on "the torus with no holes"—in other words, the sphere:

$$H(0) = [½(7 + \sqrt{1})] = [4] = 4.$$

Unfortunately, we cannot deduce the four-color theorem from the proof of the Heawood conjecture, which applies only when h is a positive number.

HEINRICH HEESCH

In the 1960s, much of the work on the four-color problem was still piecemeal. Attempts to find unavoidable sets and reducible configurations were largely independent of one another, and a coordinated search for the holy grail of an unavoidable set of reducible configurations had yet to meet with any success. The credit for advocating such a search goes to Heinrich Heesch, whose contributions would pave the way for the eventual solution of the problem by Kenneth Appel and Wolfgang Haken in the 1970s.

Heesch's early career in mathematics was spectacular. After graduating in both mathematics and music from Munich, he obtained a doctorate in mathematics from the University of Zurich for a dissertation on problems related to symmetry in three-dimensional geometry. He then moved to Göttingen, the home of such distinguished mathematicians as David Hilbert, Richard Courant, and Hermann Weyl, where he became Weyl's assistant, studying the geometry of crystals. While in Göttingen, he gained some notoriety for contributing to the solution of one of the celebrated *Hilbert problems*.

Heinrich Heesch (1906–1995).

David Hilbert, arguably the greatest mathematician of his time, gave a lecture at the Second International Congress of Mathematics, held in Paris in 1900, in which he set out the twenty-three mathematical problems that he hoped to see solved during the twentieth century. Most of his problems have indeed been settled, while others remain unanswered to this day. The four-color problem, lying outside the mainstream of mathematics, was not one of these twenty-three problems, but the *regular parquet problem*, on the construction of a particular type of tiling of the plane, was part of Hilbert's Problem 18. Heesch solved this problem in 1932 by constructing a number of tilings that can be used to cover the plane according to the rules laid down in the problem. One of his tiling patterns was later incorporated into the ceiling of Göttingen's library.

The mid-1930s were difficult years for Heesch, as they were for many other German academics and intellectuals. He was unsympathetic to-

Heesch tiling.   The ceiling of
                     Göttingen's library.

wards the National Socialists, objecting in particular to the Nazi work camps that were set up for aspiring professors, and soon found himself without university employment. For twenty years Heesch taught mathematics and music in various schools and worked on industrial problems that involved tiling patterns while continuing with his researches into mathematics. He eventually secured a teaching position at the Technical University of Hanover.

Heesch's interest in the four-color problem dates from around 1935. One of his friends, Ernst Witt, believed that he had found a solution to the four-color problem, and the two men visited Richard Courant to show him the solution. Courant was about to leave for a train journey to Berlin, so Heesch and Witt bought train tickets and accompanied him on his journey. Failing to convince Courant that the proof was correct and feeling rather deflated, they caught the train back to Göttingen; on the return journey, Heesch found an error in Witt's solution.

While he was working on the four-color problem, Heesch gradually came to believe that the right approach was to seek an unavoidable set of reducible configurations. But he feared that such a set would be very large, possibly containing up to ten thousand configurations.

In order to produce unavoidable sets, he invented the method of discharging, which we outlined in Chapter 8. He also developed the uncanny knack of being able to look at a configuration and say whether it was reducible—with at least 80 percent accuracy. As Wolfgang Haken later remarked:

> What fascinated me most was that Heesch looked at the configuration, and he either said, 'No, there is no chance. That cannot be reducible', or he said, 'But this one: that is certainly reducible'. And I asked him, 'How do you know? How can you tell?' 'Well, I need two hours' computer time . . .'

WOLFGANG HAKEN

In the late 1940s, Heesch presented his findings publicly for the first time in lectures at the University of Hamburg and in his home town of Kiel in Schleswig-Holstein. In these lectures Heesch expounded on his belief that there exists an unavoidable set of reducible configurations, that these configurations should not be particularly large, but that there is likely to be a very large number of them.

One student who attended Heesch's 1948 lecture at the University of Kiel was the young Wolfgang Haken. Haken had entered the university as its youngest student and was studying mathematics, philosophy, and physics. Haken recalls hearing Heesch's talk, much of which he did not understand at the time, but remembers Heesch mentioning that there might be some ten thousand cases to be investigated and that five hundred configurations had already been checked—at an average rate of about one per day. Heesch seemed optimistic about dealing with the remaining nine-and-a-half thousand.

Some of the most stimulating lectures that Haken attended at Kiel were those on topology given by the only mathematics professor, Karl Heinrich Weise, who had worked on the four-color problem. In these lectures Weise described three long-standing unsolved problems. First was the *knot problem*, of determining whether a given tangle of string in three dimensions contains a knot—for example, the first tangle below can be unknotted, while the second one must always remain knotted.

unknotting a tangle   knotted tangle

The second problem was the *Poincaré conjecture*, which concerned the classification of spheres in four-dimensional space; this conjecture remained unproved until 2002, when it was successfully dispatched by Grigory Perelman. The third was the four-color problem.

Like David facing Goliath, Haken took on all three problems, in each case using "very elementary means on which everyone else had given up". The first Goliath he slew. The second he attacked vigorously, rendering the giant 99 percent vanquished, but his victim recovered. The third Goliath he battled with for several years, eventually replacing the stone in his sling by an appel, as we shall see in Chapter 10.

In his doctoral dissertation, written under the supervision of Professor Weise, Haken worked on three-dimensional topology and obtained some partial results on the knot problem. Subsequently he succeeded in solving the problem in full—a great achievement—and announced his results at the 1954 International Congress of Mathematicians in Amsterdam, where he was urged to write up the details of the proof. But by this time, unable to secure a university position in Kiel or elsewhere, he was working in Munich as a physicist for Siemens, the electrical engineering company. It took Haken four years to find the time to write out the two hundred pages of his proof in intricate detail, and it was to be tablished the *D*-reducibility of many more configurations of increasing complexity.

The complexity of a configuration is measured by its *ring size*, which is the number of countries surrounding the configuration. For the Birkhoff diamond, which has ring size 6, there are thirty-one essentially different colorings of the countries in the surrounding ring of six countries, as we saw in Chapter 8.

Birkhoff diamond

Unfortunately, the number of different colorings increases rapidly as the ring size increases:

ring size	6	7	8	9	10	11	12	13	14
colorings	31	91	274	820	2461	7381	22,144	64,430	199,291

For example, the configuration of eight countries shown below has ring size 14, and we have to consider no fewer than 199,291 essentially different colorings of the countries in the surrounding ring. Informal calculations indicated that to solve the four-color problem might require the examination of configurations with ring size 18, for which the number of colorings of the surrounding ring exceeds sixteen million.

Heesch and Dürre found that the time taken to analyze a configuration increased rapidly as the ring size increased. On their computer a typical configuration with ring size 12 might take six hours to analyse, while some configurations with ring size 13 took anything between 16 and 61 hours, and those of ring size 14 were way over the horizon.

Indeed, they estimated that to verify all 10,000 cases, they might need anything from 3000 to 50,000 hours of computer time, which was not a realistic proposition on the Hanover computer—or indeed on any computer of the time.

One useful simplification that Dürre discovered was a more efficient way of storing the colorings of the surrounding ring. For a configuration of ring size 14, storing the colors of the fourteen countries in the surrounding ring would require twenty-eight bits, using two bits to specify the color (00, 01, 10, or 11) of each country. Since there are almost two hundred thousand possible colorings for a configuration of ring size 14, more than five million bits would be needed for each such configuration. Dürre developed a brilliant technique that enabled him to assign a single bit to each coloring, thereby reducing the total amount of computer time for each configuration by a factor of twenty-eight—a considerable help. Although most configurations of ring size 14 were still too large to be handled, many smaller configurations could now be checked for the first time.

By this time, mathematicians could see that any solution of the four-color problem along these lines would be complicated. In the 1960s, Edward F. Moore of the University of Wisconsin developed the remarkable knack of constructing large and intricate maps that contained no known reducible configurations, thereby hoping to find a map that required five colors. The following diagram shows part of a map he devised that contains no reducible configurations with ring size 11 or less: here the left-hand and right-hand sides of the map must be joined up, so that the top and bottom countries are both nine-sided. The existence of this map shows that any unavoidable set of reducible configurations must necessarily contain at least one reducible configuration of ring size 12 or greater.

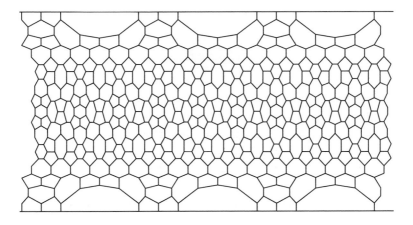

Although it looked as if configurations of ring size up to 18 might be necessary, Appel and Haken's solution would be based on configurations with ring size 14 or less. Had it required configurations of ring size 15, Appel and Haken would almost certainly not have been able to complete their solution when they did.

By this time, almost everyone working on the four-color problem was using the dual formulation, introduced by Kempe, of coloring the points of the corresponding graph or linkage. (Unlike Heesch and his successors, we avoid a sudden change of gear at this point by continuing to present our story in terms of maps.) In particular, Heesch devised a useful notation, which soon became widely used, for representing each point by an appropriate "blob" to make it easily distinguishable. Here are four of Heesch's symbols followed by an example of a Heesch drawing and a corresponding configuration of countries.

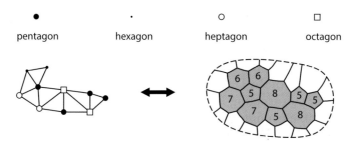

Although Heesch lectured widely on his work, it remained unpublished until 1969, when he produced a paperback book in German setting out the method of discharging and his many other contributions. Oystein Ore, in his 1967 book on the four-color problem, had made no mention of Heesch's work, of which he was presumably unaware. However, Heesch was pleased to learn about several new reducible configurations from Ore's book.

COLORING HORSESHOES

It soon became clear that the Hanover computer was not powerful enough to carry out the work required of it. Haken tried to help Heesch and Dürre by bidding for time on the large new parallel supercomputer at the University of Illinois, the ILLIAC-IV, whose construction was nearing completion, but it was not yet ready for use. Eventually, John Pasta, head of the university's computer science department, referred

Heesch and Dürre to his friend Yoshio Shimamoto, Director of the Computer Center at the Atomic Energy Commission's Brookhaven Laboratory in Upton, Long Island. At this laboratory there was a Control Data 6600, the most powerful computer of its day.

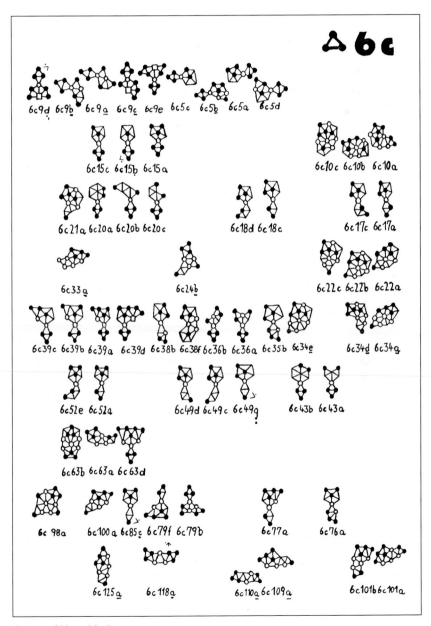

A page of Heesch's diagrams.

Fortunately, Shimamoto had always been a devotee of the four-color problem, and as director he was allowed to use up to 10 percent of his time on the computer for his own purposes. He was strongly supportive of Heesch's approach to the problem and generously invited Heesch and Dürre to visit Brookhaven and continue their reducibility testing on the Cray machine. Heesch paid two extended visits, including a stay of one year, while Dürre remained there for almost two years, completing a doctoral thesis on reducibility at the same time.

At Brookhaven, Dürre first had to convert his programs from Algol to Fortran, but once this was done progress was quickly made. A configuration with ring size 13, with 66,430 colorings to check, was not excessively large for the computer, and configurations with ring size 14 could be tested for the first time. Eventually, Heesch and Dürre were able to confirm the D-reducibility of more than a thousand configurations of ring size 14 or less.

Early in the process, Dürre discovered a small bug in the program, which meant that the first few configurations on the list were not ab-

Heinrich Heesch, Yoshio Shimamoto, and Karl Dürre at Brookhaven National Laboratory.

solutely reliable. Since each of these configurations would have taken several hours to rework, he postponed the task. In the meantime, he repaired the bug, leaving strong warnings about these early configurations—but these warnings were lost.

Meanwhile, Shimamoto was pursuing his own researches on the four-color problem, using a different but related approach. Unlikely as it may seem, he was able to show that if he could find a single configuration with certain properties, and if this configuration were D-reducible, then the four-color theorem would follow: the whole of the proof would depend on a single configuration!

Suddenly, on September 30, 1971, he found what he was seeking. Like the Bishop of London in the 1880s (see Chapter 6), Shimamoto was attending a rather boring meeting and he started to color maps—eventually producing the following configuration with ring size 14 that became known as the *Shimamoto horseshoe* (even though it looks nothing like a horseshoe). Shimamoto had proved that if this horseshoe were D-reducible, then the four-color problem was solved.

Shimamoto horseshoe

After the meeting, Shimamoto bumped into Heesch and Haken, who were visiting Brookhaven and were on their way to the cafeteria for lunch. Heesch recognized the horseshoe as one of his list of D-reducible configurations. Shimamoto, although obviously excited, was cautious and asked for the configuration to be rechecked for D-reducibility. Dürre came over from Germany to do the rechecking, but the original printouts were no longer available and so the whole task had to be run from scratch.

Excitement ran riot! Rumors about Shimamoto's solution were already spreading around the world. The day after the horseshoe's discovery, Haken was due to visit the Princeton Institute for Advanced Study before returning to Illinois, and he asked Shimamoto if he could to tell people about the result. According to Haken, Shimamoto said *yes*, while

Shimamoto later denied doing so and became rather annoyed about Haken's pronouncement, complaining that

> I am getting a deluge of questions from many people and I am now wondering about the wisdom of having you talk at Princeton and Urbana before a clean manuscript had been prepared.

In the event, things did not work out as hoped. In the first computer run, the time taken exceeded the original run time by more than an hour, and it was duly terminated. The second run took even longer. Finally, in a third attempt, the superfast computer was allowed to run for a whole weekend, and after grinding on for twenty-six excruciating hours, it confirmed that the horseshoe was, after all, *not D-reducible*. Whether the horseshoe was one of Dürre's bugged configurations was unclear, but there was certainly a sense of great disappointment all around.

Haken reworked Shimamoto's theoretical arguments and found them completely correct. Hassler Whitney and Bill Tutte, the most distinguished graph theorists of their day, also analyzed the Shimamoto approach in detail and wrote a lengthy and influential paper that began:

> In October 1971 the combinatorial world was swept by the rumour that the notorious Four Colour Problem had at last been solved . . . with the help of a computer.

They then summarized the method:

> Shimamoto, on the assumption that the Four Colour Conjecture was false, showed that there must be a non-colourable map M containing a configuration H [the horseshoe] that had already passed the computer test for D-reducibility. He then arrived at a contradiction by showing that the D-reducibility of H implied the 4-colourability of M . . . The burden of proof was not now on a few pages of close reasoning, but on a computer!

All along, Whitney and Tutte had felt that, if Shimamoto's proof were valid, a much simpler proof should be achievable by confining one's attention to a smaller part of M and that "this simpler proof would be so simple that its existence was incredible". Since they found no essential flaw in Shimamoto's reasoning, they deduced that the computer result must be wrong. Moreover, they proved that not only was the horseshoe not D-reducible, but so also was *any* configuration produced by Shimamoto's method.

It seemed as though the four-color problem had once again come to a dead end.

10

Success!

In spite of the horseshoe episode, there were reasons for optimism. Heesch's approach to the four-color problem was beginning to pay dividends, and in the hands of Appel and Haken would yield the eagerly desired goal of an unavoidable set of reducible configurations within the next five years.

In Chapter 8 we saw that Heinrich Heesch's method of discharging involved assigning an electrical charge of $6 - k$ to each country bounded by k edges, so that each pentagon receives a charge of 1, hexagons receive zero charge, and heptagons, octagons, etc., receive a negative charge. In 1970, Heesch sent Haken the results of a new discharging experiment in which the positive charge on each pentagon was now to be distributed *equally* among all neighboring countries with negative charge. The result of this experiment, if applied to a general map, would yield about 8900 "bad" configurations extending up to ring size 18, in which some countries would still have positive charge. This approach, called the *finitization of the four-color problem*, reduced the problem to the consideration of these 8900 configurations, which Heesch proposed to work through one at a time.

Wolfgang Haken, however, was deeply pessimistic about having to deal with such a large number of configurations, especially since some of them were fairly big. By this time it had become a simple matter to test configurations of ring size 11, since there are only 7381 different colorings of the surrounding ring of eleven countries, but each time the ring size increased by one, the amount of computer time required seemed to expand by a factor of four, with a corresponding increase in storage requirement. Recalling that the difficult horseshoe configuration with ring size 14 had taken many hours to run, Appel and Haken later observed:

> Even if the average time required for examining 14-ring configurations was only twenty-five minutes, the factor of 4 to the fourth power in passing from 14- to 18-rings would imply that the average eighteen-ring configuration would require over a hundred hours of time and much more storage than

was available on any existing computer. If there were a thousand configurations of ring-size 18, then the whole process would take over a hundred thousand hours, or about eleven years, on a fast computer.

A HEESCH-HAKEN PARTNERSHIP?

For some time, Haken had felt that the complexity of the problem would be substantially reduced if a better discharging method could be found. By focusing on maps without hexagons or heptagons, he succeeded in finding a much simpler procedure. Encouraged by this, he moved on to maps in general and sent some of his results to Heesch. Heesch was impressed and invited Haken to collaborate with him.

In 1971 Heesch sent Haken several of his unpublished results on reducible configurations. These included three *obstacles to reducibility*, which seemed to prevent a configuration from being reducible. Although it has never been proved that a configuration containing one of these obstacles cannot be reducible, no such reducible configuration has ever been found, so it seemed sensible to exclude them from consideration. The general study of such obstacles was soon developed by Whitney and Tutte in their paper on Shimamoto's work and later by Walter Stromquist, a graduate student at Harvard University, who showed that Heesch's obstacles are the three most important. Stromquist would soon make a name for himself in the map-coloring world by proving that the four-color theorem is true for all maps with up to fifty-one countries.

Heesch's three obstacles are these:

    4–legger country   3–legger articulation country   hanging 5–5 pair

The first is a *four-legger country*—a country C that adjoins four consecutive countries (marked with stars) of the surrounding ring. The second is a *three-legger articulation country*, a country C that adjoins three countries of the surrounding ring that are not mutual neighbors. The third is a *hanging 5–5 pair*, a pair of adjacent pentagons that adjoin a single country C inside the surrounding ring.

But by this time, Haken was beginning to change his approach to the problem. Unlike everyone else, whose objective seemed to be to collect reducible configurations by the hundreds and then package them up into an unavoidable set, Haken's main line of attack (which he later developed with Appel) was to aim directly for an unavoidable set. This set was to contain only configurations that were *likely* to be reducible—in particular, they should contain none of the reduction obstacles—in order to avoid wasting time checking configurations that would eventually be of no interest. Any configurations that subsequently proved not to be reducible could then be dealt with individually. As Haken later commented:

> If you want to improve something, you should not improve that part which is already in good shape. The weakest point is always the one you should improve. This is a very simple answer to why we got it and not the others.

He clearly felt that it was inappropriate to spend a lot of expensive computer time in checking the reducibility of configurations that would be unlikely to appear in the eventual unavoidable set.

Thus, from this point on, Haken headed off in a different direction from everyone else, concentrating on the unavoidable set and leaving the details of reducibility testing until much later. Heesch, while initially sympathetic to these ideas, soon came to reject the idea of a "likely-to-be-reducible" configuration. Their working relationship must have been further damaged by the horseshoe episode, and Shimamoto had understandably decided that he no longer wished to cooperate.

Back in Germany, Heesch was having immense problems obtaining the necessary funding for the use of a powerful computer. As a relatively minor figure in the academic hierarchy, he had little clout, and his proposal for funding did not receive the consideration it deserved. The unsympathetic grant referee apparently showed little understanding of the problem or of Heesch's approach to it, causing Shimamoto to protest:

> The remarks made by this referee are so inane that they hardly deserve any attention. It is evident that he knows nothing about the problem, and by the bias shown in his comments, he should disqualify himself from reviewing proposals involving the four-color problem. It comes to me as something of a surprise that the Foundation itself did not see it appropriate to ignore his comments.

KENNETH APPEL

Meanwhile, Haken, with little knowledge of computing and with the Brookhaven machine no longer available, also considered setting the problem aside for a few years until more powerful computers became available to deal with the massive calculations that would clearly be necessary. He had talked to computer "experts", who told him that his ideas could not be programmed, and during a lecture that he gave in Illinois on the horseshoe episode he exclaimed:

> The computer experts have told me that it is not possible to go on like that. But right now I'm quitting. I consider this to be the point beyond which one cannot go without a computer.

Present at this lecture was Kenneth Appel. Appel had graduated from Queens' College, New York, before receiving his doctorate from the University of Michigan for a dissertation on the application of mathematical logic to some problems in algebra. He was also an experienced computer programmer: he had learned programming at the University of Michigan and gained more computing experience during a summer placement with Douglas Aircraft. After working at the Institute for Defense Analysis at Princeton for two years, he settled at the University of Illinois at Urbana-Champaign. Appel's computer skills were to prove invaluable in the solution of the four-color problem.

After the lecture, Appel told Haken that he thought the experts were talking nonsense—their opinion probably reflected their unwillingness to invest a great deal of time in something whose outcome was uncertain. Appel offered to work on the problem of implementing the discharging procedures, saying: "I don't know of anything involving computers that can't be done; some things just take longer than others. Why don't we take a shot at it?" Coincidentally, Thomas Osgood, a research student of Haken's, had just submitted his doctoral thesis on solving the four-color problem for maps containing only pentagons, hexagons, and octagons. Appel was one of the members of Osgood's thesis examination panel, so the collaboration might well prove beneficial to all concerned.

Haken was delighted to accept Appel's offer to take care of the computing side of things. They decided to concentrate their search on unavoidable sets, without taking time to check the configurations for reducibility. Instead, they focused on *geographically good configurations*— configurations that contain neither of Heesch's first two obstacles to reducibility, the four-legger country and the three-legger articulation country: such configurations can easily be identified by a computer—or, indeed, by hand. They would then check their configurations for reducibility once the entire set had been constructed.

Another useful rule of thumb they used to identify suitable configurations for inclusion was the *m-and-n rule*: if n is the ring size of a configuration without obstacles and m is the number of countries inside the surrounding ring, then the likelihood of reducibility depends on the relative sizes of n and m; in particular, if m is larger than $(\frac{3}{2} \times n) - 6$, then the configuration is almost certainly reducible. For example, the reducible Birkhoff diamond has ring size 6 and 4 interior countries, and 4 is larger than $(\frac{3}{2} \times 6) - 6 = 3$; on the other hand, the Shimamoto horseshoe, which is not reducible, has ring size 14 and 10 interior countries, and 10 is smaller than $(\frac{3}{2} \times 14) - 6 = 15$.

GETTING DOWN TO BUSINESS

When they started work in late 1972, Appel and Haken had no clear idea of where they were heading. As Appel recalls, "We started with certain ideas and kept discovering that we had to become more sophisticated to avoid being swamped by useless or repetitive data". However, their first exploratory computer runs were already providing much useful information—in particular, that geographically good configurations of reasonable size (with ring size 16 or less) tended to appear near most countries that ended up (after discharging) with positive charge. However, the computer output was enormous, with many configurations appearing over and over again. It would be necessary to keep these duplications under control if the eventual list was to be manageable. Fortunately, since the computer runs had taken just a few hours, Appel and Haken realized that they could experiment as frequently as they needed to.

The necessary changes to the program were straightforward to implement, and the second set of runs, a month later, saw a definite improvement. The printout was also much reduced in thickness and would eventually be down to a fraction of an inch. But as the larger problems were overcome, lesser ones began to emerge.

From then on, every two weeks or so they modified the discharging algorithm or the computer program so that the program grew while the output shrank. Their two-way dialogue with the computer continued as each successive problem was sorted out and new ones arose. Six months of experimenting and improving their procedures showed

them that their method for producing a finite unavoidable set of geographically good configurations in a reasonable amount of time was
indeed feasible.

At this stage, they changed tack. They decided to prove *theoretically*
that their method would provide such an unavoidable set. To do this
they would have to include every possible case, even if it was unlikely to
occur in practice. This proved to be much more complicated than they
had expected, taking them more than a year to complete. But the eventual outcome, in the autumn of 1974, was a lengthy proof that an unavoidable set of geographically good configurations exists, together
with an achievable method for constructing such a set. Shortly after-
wards, Walter Stromquist, as part of his Harvard doctoral thesis, produced a shorter and more elegant proof of the existence of such a set.

Appel and Haken's next task was to decide how complicated the entire process would be. They decided to experiment with a particular
case—maps that do not contain two adjacent pentagons. This was much
simpler than the general case and produced a set of just 47 geographically good reducible configurations of ring size 16 or less, as shown
opposite. This experiment enabled them to estimate that the general
problem might be only about fifty times as bad as this, and on this basis
they decided to proceed. In the event, their estimate turned out to be
overly optimistic.

In early 1975 they introduced the third of Heesch's obstacles, the
hanging 5–5 pair. Inevitably, this necessitated further changes in procedure, but it was carried out successfully with only a doubling in the size
of the unavoidable set. They also programmed the computer to search
for sets of configurations with relatively small ring size.

At this stage the computer started to think for itself, as Appel and
Haken later recalled:

> It would work out complex strategies based on all the tricks it had been
> 'taught' and often these approaches were far more clever than those we
> would have tried. Thus it began to teach us things about how to proceed that
> we never expected. In a sense it had surpassed its creators in some aspects of
> the 'intellectual' as well as the mechanical parts of the task.

As soon as it seemed probable that Appel and Haken would be able
to find an obstacle-free unavoidable set of configurations, which were
therefore likely to be reducible, it was time to start the massive detailed
check for reducibility. Inevitably, there would be some rogue reducible
configurations in the list, but they hoped that these would be relatively
few in number. Also, with configurations that might go up to ring size

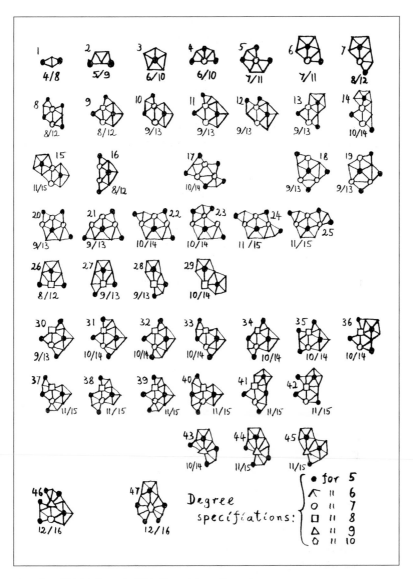

Appel and Haken's geographically good configurations.

16 and others that might cause trouble in other ways, they were expecting to have to devise some shortcuts.

By the middle of 1974, Appel and Haken knew that they would need help with the reducibility programs. Appel visited the University of Illinois computer science department to ask whether any graduate student would be interested in writing a thesis in the area of programming.

Coincidentally, John Koch had just been scooped by a journal article that solved the thesis problem on which he had been working for a year and was looking around for another topic. Koch's thesis project was set up so that its completion would not depend on the outcome of Appel and Haken's attack on the four-color problem.

Koch was put to work on the C-reducibility of configurations of ring size 11. As we saw with the Birkhoff diamond in Chapter 8, a C-reducible configuration is one that can be adapted so that the reducibility arguments go through more easily, but it is not always clear how this adaptation can be carried out. Appel and Haken were particularly interested in two types of modification that were relatively easy to implement, and Koch discovered that 90 percent of the configurations of ring size 11 were of these types. Agreeing that little would be gained by including the other 10 percent, which would have required some difficult programming, they decided to concentrate exclusively on the simple modifications. Koch devised an elegant and highly efficient method for testing the C-reducibility of these configurations of ring size 11, and Appel subsequently extended it to configurations of ring sizes 12, 13, and 14.

By the end of 1975, their work on the discharging method had run into trouble. Structural changes would be necessary, requiring significant modifications to the program. The problem was that, while trying

John Koch and his wife, Mimi, with his doctoral certificate.

Kenneth Appel and Wolfgang Haken contemplate their solution.

to disperse the positive charge on each pentagon to its immediate neighbors, they regularly came up against barriers of hexagons, all with zero charge. For Wolfgang Haken, taking a vacation is doing mathematics twenty-three hours a day in a different location, and while he was walking on the beach during his annual family holiday at Key West in Florida, he asked himself why the positive charge on the pentagons should not be permitted to "jump over" these hexagon barriers. This would make for a more efficient process but would leave Appel and Haken with a dilemma: should they rewrite their program from scratch, or should they adopt a more ad hoc approach? Since the former would have been a horrendous task, they opted for the latter, implementing the final version of the discharging process by hand. This would inevitably require much work, but it would give them extra flexibility by enabling them to make minor changes whenever necessary. As it turned out, this led to so many improvements that they could restrict all their configurations to ring size 14 or less.

THE FINAL ONSLAUGHT

Throughout the first half of 1976, Appel and Haken worked on the final details of the discharging procedure that would give them their unavoidable set of reducible configurations. To do this, they sought out problem configurations that would necessitate further changes to the discharging procedure. Whenever they found such a configuration, they immediately tested it for reducibility—which could usually be done fairly quickly—and in this way, the reducibility testing by computer could keep pace with the manual construction of the discharging procedure. In the event, the final process used 487 discharging rules, requiring the investigation by hand of about ten thousand neighborhoods of countries with positive charge, and the reducibility testing by computer of some two thousand configurations.

Reducible configurations seemed to be everywhere, like trees in a forest, but Appel was never quite sure . . .

> if you shoot a gun in any direction it's eventually going to hit a tree . . . there's a tree there no matter where you look. But yet you haven't got enough of a description of a forest to formally prove there is a tree there, and the thing you always worry about is there is one little direction where you just shoot it and there is no tree, and it goes out of the forest entirely . . . I just think of that—being in a forest and shooting a gun.

Because the reducibility of an awkward configuration could sometimes take a long time to check, and with memories of the nonreducible Shimamoto horseshoe (which took twenty-six hours), they found it convenient to impose on each configuration an artificial limit of ninety minutes' checking time on an IBM 370–158 computer or thirty minutes on an IBM 370–168 computer. If a configuration could not be proved reducible in this time, it was abandoned and replaced by other configurations; finding such alternative configurations was usually straightforward. By way of comparison, they estimated that to check the computer output of one of the more difficult configurations in full detail would take someone working a forty-hour week about five years in total.

The last few months were indeed extremely heavy on computer time, but here Appel, Haken, and Koch were very fortunate. Probably no other institution would have allowed them twelve hundred hours of computer time, especially since the outcome of the whole process was still uncertain. But the University of Illinois computer center was very supportive throughout, and included the four-color team among a small group of computer users who were assigned spare computer time when the facility was not being used for anything else. Use was also made of the computer at the university's Chicago campus: programs were sent to Chicago overnight, and the returns were ready by the next morning.

In March 1976, the university's administration unit acquired a powerful new computer. As Haken recalls, local politicians started to ask, "Why does this administration need a larger computer than the scientists"? and the administrators had to promise, "OK, half the time, or whenever the thing is not needed by us, the scientists can use it". Since Appel seemed to be the one scientist who could get the machine to run properly, he had almost exclusive use of it at first, and he benefited from a valuable fifty hours of computer time over the Easter vacation. Everyone was happy: the administrators could claim a more balanced loading of the system over a twenty-four-hour period, while Appel got all the computer time he needed.

In the event, the new computer proved to be so powerful that everything proceeded far more quickly than they had expected, saving Appel and Haken, by their own estimation, a full two years on the reducibility testing. Meanwhile, with the help of Haken's daughter Dorothea, they began an extremely exhausting and stressful few months working in detail through the two thousand or so configurations that would eventually form the unavoidable set:

> one of us wrote a section, and then the second one was reading it over, checking it for mistakes, and bringing the errors to the attention of the one who wrote it, and then the third person read it a third time. Reading it over was more strenuous than writing it, because it went faster, and it was more exhausting. It was a terrific amount of work for three people.

Suddenly, by late June, almost before they realized what was happening, the entire job was finished. The Hakens had completed the construction of the unavoidable set, and within two days Appel was able to test the final configuration for reducibility. He celebrated their achievement by placing a notice on the department's blackboard:

> **_Modulo careful checking,_**
> **_it appears that_**
> **_four colors suffice_**

This phrase *four colors suffice* subsequently became the mathematics department's postal meter slogan:

A RACE AGAINST TIME

All that remained was the detailed checking, which had to be done quickly. Not expecting the reducibility testing to advance so well, Appel had arranged a sabbatical visit to Montpellier, where he would be working with Jean Mayer, a professor of French literature with a passion for map coloring. Appel was due to leave in late July. "We have five weeks", he said. "Either we announce our result in five weeks, or we have to wait five months".

Time was indeed of the essence: unknown to them, several other map colorers were close to solving the problem. At the University of Waterloo in Ontario, Frank Allaire had the best reducibility methods around. These were inherited from Jean Mayer and were, in Haken's words, "even better than Heesch's and much better than ours". Appel and Haken were aware of Allaire's superior methods, and on one occasion Appel had written to him for confirmation that an awkward configuration was indeed reducible; this was the only occasion on which they asked for Allaire's help. By 1976, Allaire was several months ahead of them in his investigations into reducibility and was expecting to complete his solution within a few months.

Meanwhile, at the University of Rhodesia (now Zimbabwe) was a former chemist, Ted Swart, who had carried out the first radio-carbon dating in Africa. He was working independently on the four-color problem and was making excellent progress. Swart submitted a paper to the *Journal of Combinatorial Theory* and received a reply from the editor-in-chief, Bill Tutte (also at Waterloo), to the effect that Allaire was working along very similar lines. Allaire and Swart pooled their results and submitted a paper just before the Appel-Haken proof was announced. This paper described an algorithm for determining the reducibility of configurations and included a full list of all reducible configurations with ring size 10 or less. And there were other rivals. The Harvard doctoral student Walter Stromquist had been developing some powerful new

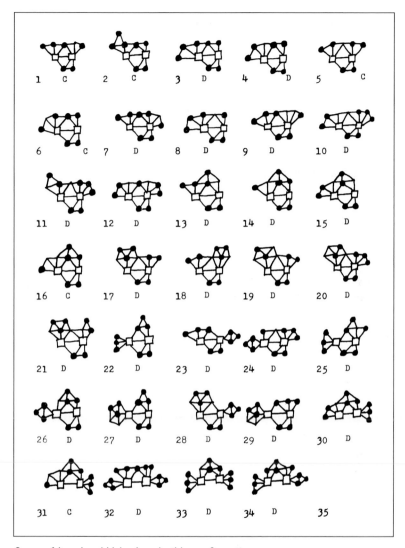

Some of Appel and Haken's reducible configurations.

methods for tackling the four-color problem and was expected to complete his solution within a year, while Frank Bernhart (whom we met briefly in Chapter 8) was also contributing some impressive work on reducibility arguments.

Although no one was aware just how close everyone else was to completing their solutions, Appel and Haken suspected that it would be too risky to wait five months, especially if a rumor were to leak out that they had almost reached their goal. They guessed that Allaire, with his superior reducibility methods, was some way ahead of them on that score but were confident that their strategy of spending nine-tenths of their time on unavoidable sets and only one-tenth on reducible configurations, while others operated the other way around, gave them an edge.

Even so, they had no time to lose. Enlisting five of their children to help—Dorothea and Armin Haken and Laurel, Peter, and Andrew Appel—they immediately set to work. They had high expectations that, though the checking might turn up many typos and the occasional bad configuration that would need replacing, there would be no major disasters.

Laurel Appel checked through the seven hundred pages of detailed working and found about one error per page, most of them typos; all but fifty of these she immediately corrected herself. Appel then spent the Fourth of July weekend recomputing these fifty cases, and only twelve of them now failed to work. These twelve were replaced by twenty new ones, which were then reduced to just two, and so the process continued. As Haken later remarked:

> Somebody has worked one month full-time and found eight hundred mistakes. And then we needed only five days to repair all that. This looks so stable—it is this incredible stability . . .

Appel and Haken knew by this stage that they were safe: even if a few configurations turned out not to be reducible after all, there was more than enough self-correction in the whole system for these configurations to be replaceable easily and quickly. It was not possible for a single faulty configuration, if one existed, to cause the entire edifice to collapse. What is more, since they could adapt their list to produce many hundreds of unavoidable sets of reducible configurations, they had not just one proof, but hundreds of proofs, of the four-color theorem!

By now very confident, they decided to go public. On July 22, 1976, just a few days before Appel was due to leave for France, they formally told their colleagues and sent out complete preprints to everyone in the field who might be approached by other mathematicians or the press for comment. The last thing they wanted was a grumpy "Well, no one told me" kind of response.

One recipient of a preprint was Bill Tutte. Two years earlier he had written an article in the *American Scientist* claiming that people who used their approach were real optimists, since the method had then seemed extremely unlikely to work. But when Tutte heard the news he waxed eloquent, comparing their achievement with the slaying of a fabled Norwegian sea monster:

> Wolfgang Haken
> Smote the Kraken
> One! Two! Three! Four!
> Quoth he: 'The monster is no more.'

And when Tutte was interviewed by the press, he told them, "If they say they've done it, I have no doubt that they've done it".

Appel and Haken were delighted that a mathematician of Tutte's stature should have given his positive support so quickly. Tutte's endorsement would go a long way to setting people's minds at rest, whereas a lukewarm response could have raised serious doubts about their solution.

AFTERMATH

After the blackboard announcement in early July, Andrew Ortoni, a stringer for *The Times* of London who worked at the University of Illinois, asked Appel and Haken's permission to break the news. They asked him to wait until the checking was complete but promised to let him know as soon as they were ready to announce. On July 23, 1976, the following report duly appeared in *The Times*:

> Two American mathematicians have just announced that they have solved a proposition that has been puzzling their kind for more than 100 years . . . Their proof, published today, runs to 100 pages of summary, 100 pages of detail and a further 700 pages of back-up work. It took each of them about 40 hours research a week and 1000 hours of computer time. Their proof contains 10,000 diagrams, and the computer printout stands four feet high on the floor.

Other papers around the world, from the Japanese *Asahi Shin-bun* to the German *Neue Zürcher Zeitung*, latched on to the story, and there was great excitement. The solution was featured in *Time* and *Scientific American*, and some of the configurations appeared on the front cover of *New Scientist*. Only the *New York Times* seemed to have a policy of not publishing announcements of proofs of the four-color theorem, "because they were all false anyway". In September, according to Appel, Lipman Bers, president of the American Mathematical Society, approached the paper, saying:

> 'Look, you know, you never have published this thing. It's pretty well accepted.' And they said, 'Well, it's too late for us to publish a news story.' And they said,

A Japanese newspaper report on the Appel-Haken solution.

'Would you write it?' And so Bers wrote an editorial in the *Times* which acknowledged it and congratulated us.

The American Mathematical Society also published a two-page research announcement by Appel and Haken in the September 1976 issue of their *Bulletin*, outlining the main ideas of the proof.

Appel and Haken decided to submit the full solution to the *Illinois Journal of Mathematics*. This was not because they felt that they had a better chance of publishing it locally, for if their proof were correct, then publication was not in doubt. A few years earlier they had published a long and turgid article in its pages, and they felt that they owed it to the *Illinois Journal* to give it the kudos for publishing their solution of this famous problem. But their main reason for favoring local publication was because they wanted to be in a position to suggest appropriate referees for their proof. It was clearly in everyone's interest—theirs and the *Journal's*—that a paper such as this should be thoroughly refereed:

> We said to the *Illinois Journal*, 'Look, if this thing is wrong, we want to know it as badly as anyone else does, and you're not going to reject it on the grounds of triviality.'

They agreed with the *Journal's* editor that the best people in the world should be selected as referees—Frank Allaire for the reducibility part and Jean Mayer for the discharging arguments.

In France, Appel spent much of his sabbatical in Montpellier working painstakingly through the details of the proof with Jean Mayer, one step at a time, dealing with Mayer's queries and suggestions as they arose. On his return in December 1976, he and Haken set to work refining the details of the proof, conferring with the referees, and preparing the paper and its associated microfiche for publication.

Understandably, Frank Allaire was bitterly disappointed when Appel and Haken announced their result, as he himself was nearing a solution. Moreover, he and Ted Swart had been taking a more systematic approach, as opposed to what they regarded as the hit-and-miss methods of Appel and Haken. But he quickly suppressed his own feelings for the general good, refereeing the reducibility part conscientiously and constructively.

Allaire first compared their configurations with his own, finding a substantial number that appeared on both lists. He next checked the C-reducible configurations that were not D-reducible. For each such configuration, one can calculate the number of "good" colorings of the surrounding ring—those colorings that extend directly, or after Kempe-chain interchanges of color, to the configuration inside: for C-reducible configurations of ring size 14, about 50,000 of the 199,291 possible colorings turn out to be good colorings. Allaire took 400 of these configurations and compared his own numbers with those obtained by Appel and Haken. In every case the numbers agreed exactly.

Heinrich Heesch was also upset that Appel and Haken had got there first—not surprisingly, since their method was so closely based on his, and solving the four-color problem had been his goal for more than forty years. But later he became very cooperative, sending Haken his own list of 2669 reducible configurations, together with supporting data on the number of good configurations. In September 1977, shortly before publication, a few discrepancies were noticed, causing Appel and Haken some alarm. However, Heesch recomputed his configurations and was soon able to confirm that in every case Appel and Haken were correct.

Appel and Haken's published paper was a substantial improvement on the rough-and-ready preprint they had sent out in July 1976. In particular, they discovered that their preprint contained several repeats of configurations. They also found many instances of one configuration appearing inside another, so that in each case they could omit the larger one. By weeding out these superfluous configurations, they managed to reduce their original list of 1936 reducible configurations to the 1482 configurations of the published version; they later reduced this number still further to 1405. However, they saw no point in aiming for the "smallest possible" number if this greatly increased the total computer time. As Appel pointed out,

> If one configuration replaces twelve, but that one configuration takes two hours and the twelve take five minutes, that doesn't make sense.

Appel and Haken's solution appeared in two parts in the December 1977 issue of the *Illinois Journal of Mathematics*. Part I was "Discharging", written by the two of them: it outlined the overall strategy of their proof and described their methods of discharging for constructing the unavoidable set. Part II, "Reducibility", was written with John Koch: it described the computer implementation and listed the entire unavoidable set of reducible configurations. The two papers were supplemented by a microfiche containing 450 pages of further diagrams and detailed explanations.

Appel and Haken had at last achieved their goal: the four-color theorem was proved.

11

Is It a Proof?

The Appel-Haken proof of the four-color theorem was greeted with great enthusiasm: after 124 years, one of the most famous problems in mathematics had finally been solved.

It was also greeted with skepticism:

we've been here before so many times—remember the Shimamoto horseshoe.

It was greeted with outright rejection:

in my view such a solution does not belong to mathematical science at all.

But more than anything else, it was greeted with great disappointment:

God wouldn't let the theorem be proved by a method as terrible as that!

Too late—He already had!

COOL REACTION

In August 1976 the American Mathematical Society and the Mathematical Association of America held a joint summer meeting at the University of Toronto, and Wolfgang Haken was one of the speakers. In a report on the event, Donald Albers described the scene:

The elegant and old lecture hall was jammed with mathematicians anxious to hear Professor Haken give the proof. It seemed like the perfect setting to announce a great mathematical result. He proceeded to outline clearly the computer-assisted proof that he and his colleagues had devised. At the conclusion of his remarks, I had expected the audience to erupt with a great ovation. Instead, they responded with polite applause!

Reaction to Haken's lecture was certainly mixed. Ted Swart, who was visiting Waterloo at the time, attended the meeting and praised the quality of Haken's presentation, recalling that the applause was a little more than merely polite. Attending with him were Frank Allaire, Frank Bernhart, and Bill Tutte. As Swart recalled:

> After it was all over, the two Franks and I huddled together and agreed among ourselves that Haken et al. 'had done it'.

And many people asked them about the proof:

> What did we think? Did we think they had really done it? And I can only tell you that we were unanimous in our view that they'd done it. Unanimous.

But this was by no means the general reaction. According to Albers:

> Mathematician after mathematician expressed uneasiness with a proof in which a computer played a major role. They were bothered by the fact that more than 1000 hours of computer time had been expended in checking some 100,000 cases and often suggested (hoped?) that there might be an error buried in the hundreds of pages of computer printouts. Beyond that concern was a hope that a much shorter proof could be found.

Indeed, acceptance of the computer part of the proof seemed to be based partly on age. Haken's son Armin, by then a graduate student at the University of California at Berkeley, gave a lecture on the four-color problem and on his contribution to its solution. At the end, the audience split into two groups: the over-40s could not be convinced that a proof by computer was correct, while the under-40s could not be convinced that a proof containing 700 pages of hand calculations could be correct.

It was clear that Appel and Haken's solution of the four-color problem had created something new in mathematics. Is it a proof? And if so, how do we know that it is a proof? Albers concluded,

> it seems that the computer-assisted work of Appel, Haken and Koch on the well-known Four-Color Problem may represent a watershed in the history of mathematics. Their work has been remarkably successful in forcing us to ask What is a Proof Today?

WHAT IS A PROOF TODAY?

The concerns about the Appel–Haken proof, and particularly about the use of the computer, continued to rumble on—and still do. Can a proof be considered valid if it cannot be checked by hand? In February 1979 a philosopher, Thomas Tymoczko, set the cat among the pigeons when he wrote a paper for *The Journal of Philosophy* entitled, "The four-color problem and its philosophical significance". In this paper, and in others written around the same time, he asked whether the proof of the four-color theorem could be regarded as valid, relying as it did on extensive use of a computer. He fully accepted that Appel and Haken had *demonstrated* that four colors suffice for all maps but felt that it was not a *proof* in the sense in which mathematical proofs were usually understood.

Among the criteria Tymoczko laid down for a valid proof were that it should be both *convincing* and *surveyable*. With the former he had no quarrel:

> most mathematicians have accepted the 4CT, and none, to my knowledge, has argued against it.

It was with the latter criterion that he mainly took issue, asserting with full force and in great detail that the Appel-Haken proof is not surveyable—that is, it cannot be checked in all its detail:

> No mathematician has seen a proof of the 4CT, nor has any seen a proof that it has a proof. Moreover, it is very unlikely that any mathematician will ever see a proof of the 4CT.

Tymoczko had no problem with the discharging procedure for producing an unavoidable set. This had been applied by hand to yield the unavoidable set of configurations, and Appel and Haken had given a rigorous and surveyable proof that it did indeed yield an unavoidable set. On the other hand, no mathematician could survey the reducibility part of the proof, because the details of the method of proof are hidden inside the computer. No appeal to computers can be considered a valid method of proof—in particular, computers can make mistakes. They might malfunction or be misprogrammed, their output might be misread, or the programs might not capture the mathematical intention—the Shimamoto episode is an example of where the computer got it wrong. If such "experiments" with computers are allowed, mathematics is in danger of becoming an empirical science, as fallible as physics. Tymoczko concluded that if we accept Appel and Haken's proof as valid, then we are forced to modify our concept of "proof" to allow computer experimentation as a new method for establishing a mathematical result.

Tymoczko's article caused a storm of protest, both within the columns of *The Journal of Philosophy* and elsewhere. One person who was particularly riled by his views was Ted Swart, who believed that Tymoczko had it all wrong. Within a matter of days he had written a lengthy response, which he then submitted to *The Journal of Philosophy*, but without success:

> Whether for reasons of professional jealousy or whatever, the editor did me the discourtesy of rejecting it outright without even submitting it to referees. Fortunately I had sent a copy to Martin Gardner who responded right away and said it would be a tragedy if my excellent article was not published. He went on to suggest that I should submit it to the *American Mathematical Monthly*, indicating that he had recommended I should do so.

The *Monthly* had the article in print within weeks, and a few months later Ted Swart was awarded a Lester R. Ford Award by the Mathematical Association of America for outstanding expository writing.

In his article, Swart first pointed out that all those who worked on reducibility testing were happy with Appel, Haken, and Koch's reducibility results, because they had to a large extent been checked independently by the use of different reducibility testing programs on different computers. On the general question of using a computer, he later commented that:

> For the most part I regard computer-assisted proof as just an extension of pencil and paper. I don't think there's some great divide which says that OK, you're allowed to use pencil and paper but you're not allowed to use a computer because that changes the character of the proof. I don't see that myself. I find such an argument strange.

Indeed, if Swart had a worry about the Appel-Haken proof, it lay in the discharging part, where a computer had *not* been used:

> It is no easy matter to check that the final working list of 1482 unavoidable configurations embraces all the unavoidable configurations arising from their discharging procedure. Thus the Haken and Appel proof is subject to some degree of uncertainty, and Tymoczko is definitely incorrect when he says that no mathematician 'has argued against it'.

As mathematical proofs become longer and computers are used more widely, the whole question of surveyability has become fraught with difficulties. A computer that performs extensive but completely routine tasks may certainly be considered at least as reliable as a human who checks by hand a proof that is long and complicated or breaks down into a large number of special cases. In Ted Swart's words, "Human beings get tired, and their attention wanders, and they are all too prone to slips of various kinds . . . Computers do not get tired". To illustrate the contrast, let us look at two celebrated proofs from an area of mathematics called group theory.

In 1963 Walter Feit and John Thompson proved a remarkable theorem in group theory. Its published proof runs to more than 250 pages of concentrated argument and initially contained a number of minor errors that were subsequently corrected. Although the possibility of human error in such a lengthy proof is substantial, their proof is generally accepted by mathematicians, most of whom have never worked through it in detail: they are happy to trust all those who *have* checked it thoroughly. Much the same can be said for Andrew Wiles's proof of Fermat's last theorem, a long-standing problem in number theory. While it is possible that either proof may subsequently be found to be in error (as was Kempe's proof after eleven years), each is currently regarded as being correct.

The problem of classifying all finite simple groups, a major area of study from 1950 to 1980, provides a contrast. It was proved that these objects called finite simple groups can be classified into eighteen general types, together with twenty-six separate examples. The work amounted to many thousands of pages by hundreds of contributors, and parts of it made substantial use of computers. One of the leaders in this classification was Daniel Gorenstein, who wrote in 1979 that:

> This is an appropriate moment to add a cautionary word about the meaning of 'proof' in the present context; for it seems beyond human capacity to present a clearly-reasoned several-hundred-page argument with absolute accuracy . . . there are no guarantees—one must live with this reality.

In spite of this, most mathematicians were happy to accept the classification of finite simple groups as fully proved, even though the difficulties were still being cleared up more than twenty years after Gorenstein had advised caution. It was indeed interesting to observe the different standards tolerated by the mathematical community towards purported proofs in group theory and graph theory.

But surveyability has not been the only question hanging over the Appel-Haken proof. Several mathematicians have complained that the proof is not sufficiently transparent. Among them was Ian Stewart, the well-known writer and broadcaster on mathematics, who complained that it did not explain *why* the theorem is true—partly because it was too long for anyone to grasp all the details, and partly because it seemed to have no structure:

The answer appears as a kind of monstrous coincidence. Why is there an unavoidable set of reducible configurations? The best answer at the present time is: there just is. The proof: here it is, see for yourself. The mathematician's search for hidden structure, his pattern-binding urge, is frustrated.

Even more forceful was Daniel Cohen, who had himself worked on the four-color problem:

In the analysis of each case the program only announced whether or not the procedure terminated successfully. The entire output from the machine was a sequence of yeses. This must be distinguished from a program which produces a quantity as output which can subsequently be verified by humans as being the correct answer . . . The real thrill of mathematics is to show that as a feat of pure reasoning it can be understood why four colors suffice. Admitting the computer shenanigans of Appel and Haken to the ranks of mathematics would only leave us intellectually unfulfilled.

Geometer H.S.M. Coxeter accepted the inevitable:

It has always seemed to me a different kind of theorem from other kinds of theorems. When the computer proof has been checked by quite a number of people, and they're all satisfied with it, well then we'll have to accept it. But I think it's very unlikely that anyone can break that proof down into something that one would regard as an ordinary proof. So it's rather in a different category than all other theorems.

George Spencer-Brown, on the other hand, refused to believe that Appel and Haken had proved anything:

Nowhere in their long and often irrelevant account do they provide the evidence that would enable the reader to check what they say. It may, or may not, be 'possible' to prove the color theorem the way they claim. What is now certain is that they did not do so . . . not only is no proof to be found in what they published, but there is not anything that even begins to look like a proof. It is the most ridiculous case of 'The King's New Clothes' that has ever disgraced the history of mathematics . . .

What everyone did agree on, however, was that the Appel-Haken proof could not be described as beautiful or elegant mathematics. In his celebrated book *A Mathematician's Apology*, the English mathematician G. H. Hardy had claimed that "there is no permanent place in the world for ugly mathematics", and even Ken Appel did not demur from this in respect of his own proof:

there were people who said, 'This is terrible mathematics, because mathematics should be clean and elegant', and I would agree. It would be nicer to have clean and elegant proofs.

While everyone would like to see a noncomputer solution of the four-color theorem, such a solution along the lines of the Appel-Haken proof is almost certainly unattainable—not least because E. F. Moore's example in Chapter 9 shows that any unavoidable set of reducible configurations must include configurations with ring size 12 or more. For a noncomputer proof, new ideas are needed, and as yet such ideas have not been forthcoming.

MEANWHILE . . .

After the Toronto meeting, Wolfgang Haken embarked on a lecture tour of the United States, explaining the details of the solution. Ken Appel was doing a similar thing in Europe, giving lectures in France, where he was spending time with Jean Mayer, and four lectures at the University of Bristol, where he had spent a sabbatical year some time earlier. These lectures were informative and convincing to those who attended, but others in England took a very different view.

Just before Christmas, Spencer-Brown gave a lecture at London University's Institute of Education in which he cast doubt on the Appel-Haken solution, pointing out, with some justification, that it could not yet be regarded as a proof since its details had not been published. Spencer-Brown had himself worked on the four-color problem up to 1964, the year of his *Laws of Form*, a book on logic that his friend Bertrand Russell mentioned in his autobiography and which has been variously described as "a work of genius" and "pretentious triviality." The appearance of Appel and Haken's proof spurred him to reenter the fray, and he worked night and day toward a solution that did not rely on a computer.

His lecture seemed, to me at least, a bizarre occasion. An entry fee was charged, and media people turned up in force and drank champagne at the front of the lecture theatre while the rest of the audience sat and waited. Spencer-Brown's proposed solution involved Kempe-chain arguments with an ingenious new twist: the colors chosen were *red, blue, red-and-blue* (*purple*) and *neither-red-nor-blue* (*white*), where the color *purple* could play the role of either *red* or *blue* whenever two Kempe chains met. A video was made of the lecture: this recording was intended to be the first published proof of the theorem, but in the event so many details were omitted that it could not be counted as such. Following this event, Spencer-Brown was invited to Stanford University for several months, where his proof was scrutinized, found to be deficient, and repeatedly corrected and rescrutinized. A version of his solution can be found (in English) as an appendix to the German edition of his *Laws of Form*.

Following the appearance of their unorthodox proof, Appel and Haken were sometimes made to feel most unwelcome. The most dramatic instance was when the head of a mathematics department refused to allow them to meet his graduate students, on the following grounds:

> Since the problem had been taken care of by a totally inappropriate means, no first-rate mathematician would now work any more on it, because he would not be the first one to do it, and therefore a decent proof might be delayed indefinitely. It would certainly require a first-rate mathematician to find a satisfactory proof, and that was now impossible.

They were made to feel that they had done something very wicked and that the innocent souls of students needed to be protected from their bad influence.

That head of department may have had a point. In 1977 Frank Allaire had attended the Seventh Manitoba Conference on Numerical Mathematics and Computing and presented his proof of the four-color theorem. Because of his different discharging method and his superior methods for reducibility, his solution had required only fifty hours of computer time. But although he produced a lengthy paper for the conference proceedings containing many of the main ideas of his proof, nothing ever appeared in a refereed journal; had it appeared, it could have provided independent corroboration for the Appel-Haken proof.

Two other publications appeared in 1977. For the readers of *Scientific American*, Appel and Haken wrote "The solution of the four-color-map problem"; this is still one of the clearest descriptions of their approach. Around the same time, a second book on the four-color problem was published. *The Four-Color Problem: Assaults and Conquests*, by Thomas Saaty and Paul Kainen, was based on an award-winning paper by Saaty that had appeared in the *American Mathematical Monthly*. However, the book's passage to publication was interrupted, as its preface records:

> A funny thing happened to this book on the way to the publisher. More than two years ago, we had completed the manuscript of a book entitled *Assaults*

> on the Four-Color Conjecture . . . Our original book was oriented towards the
> variety of approaches with which the four-color conjecture (4CC for brevity)
> has been investigated. The reader can well imagine our mixed reactions to the
> announcement, in the summer of 1976, that the 4CC had been proved—by
> a computer, no less.

Saaty and Kainen retrieved their manuscript and revised it in the light of Appel and Haken's solution. When their book eventually appeared, it became a best seller.

By the early 1980s, rumors were beginning to spread that there was a major error in Appel and Haken's proof of the four-color theorem. In view of its complexity, many errors might have been expected to appear, but this did not happen. The only significant one, which probably caused the rumors, was discovered in 1981 by Ulrich Schmidt, an electrical engineering student in Aachen. The four-color problem interested him because there were parallels with checking the design of computer chips. For his Diplomarbeit (master's thesis) at the Aachen Technische Hochschule, Schmidt spent one year (the maximum permitted time) working through 40 percent of the discharging part of the Appel-Haken proof. He found one error, which took Haken about two weeks to correct, in addition to a few misprints. In 1985 a further error—a minor drawing error in one configuration—was found by S. Saeki in Japan. Other than these, and a few more misprints, no significant error has ever been found in the Appel-Haken proof.

In 1986, Appel and Haken received a letter from the editor of the *Mathematical Intelligencer*, who had heard the persistent rumours that something was wrong with the proof, and invited them to set the record straight. Welcoming such an opportunity, they replied:

> With this kind offer we cannot but comply, although it is a pity to spike a good
> rumor. Mathematical rumors do add interest and excitement to the conversa-
> tion at mathematical meetings. However, the rumors about the Four Color
> Theorem seem to be based on a misinterpretation of the results of the inde-
> pendent check of details of the proof by U. Schmidt.

The result was "The four color proof suffices", an upbeat article describing their methods in some detail and discussing how they had corrected the error that Schmidt had identified. In 1989 there appeared their last word on the subject, a hefty tome entitled *Every Planar Map is Four Colorable*, which supplied several more details of the proof, proved some related results, corrected all the errors that had been discovered, and included a printed version of all their microfiche pages.

A NEW PROOF

In 1994 Neil Robertson, Daniel Sanders, Paul Seymour, and Robin Thomas added an exciting new chapter to the four-color saga. Over the previous ten years, they had proved some spectacular results in graph theory, and they now turned to the four-color problem because it was related to other things they were working on.

They were concerned that Appel and Haken's proof was still not fully accepted and believed that there was still doubt about its validity. The main reason was not that the proof required a computer, even though a full check of the reducibility part would have involved inputting 1482 configurations by hand into the computer and a lot of programming to test each one for reducibility. It was the unavoidability arguments that caused them most concern: the discharging algorithm that Appel and Haken had carried out by hand was very complicated, and no one had ever made an independent check of it all.

After a week of working through the details of the Appel-Haken proof, they gave up. They decided that it would be much more fun, and more instructive, to pursue a proof of their own, using the same general approach that Heesch had initiated and Appel and Haken had developed. It took them a whole year to complete the work, and the resulting proof was shorter, more robust, and more systematic than its predecessor.

The unavoidable set that they obtained contains only 633 reducible configurations—they could have reduced it further, to just 591 configurations, but only at the expense of more computer time. Moreover, their discharging procedure for proving unavoidability required only 32 discharging rules, in contrast to the 487 rules used by Appel and Haken.

When discharging the map, they needed to prove an observation of Heesch involving the *second neighborhood* of a given country C—this is the collection of all those countries that are not themselves adjacent to C but are adjacent to a country that is adjacent to C. By proving Heesch's conjecture that the second neighborhood of any country with excess charge always contains a reducible configuration, Robertson and his team were able to avoid certain complications that had dogged Appel and Haken in their proof.

Because they considered computer proofs of long and complicated results to be *more* reliable than hand-checked ones, they used the computer for both the unavoidability and reducibility parts of the proof.

(What would Tymoczko have said!) What is more, all the steps in their proof can be externally verified by anyone on their home computer in just a few hours.

The improved methods employed by Robertson and his team led to a more efficient algorithm for coloring maps. Appel and Haken's approach had given rise to a *quartic algorithm for map coloring*: this means that if a map has n countries, then the running time required for coloring the map is proportional to n^4. The more efficient methods of Robertson's coworkers led to a *quadratic algorithm for map coloring*, for which the running time is proportional to n^2. The resulting improvements are apparent from the following table, which compares the running times for various values of n for a computer performing a million calculations per second:

	$n = 10$	$n = 100$	$n = 1000$
n^2	0.0001 second	0.01 second	1 second
n^4	0.01 second	1 minute, 40 seconds	more than 11½ days

Shortly after their new proof appeared, one of its authors, Robin Thomas, wrote an interesting article outlining its main ideas and emphasizing the four-color theorem's importance by describing some surprising theorems in other branches of mathematics that turn out to be equivalent to it—these include results on the algebra of three-dimensional vectors and the divisibility of integers and one relating to matrices and tensors.

INTO THE NEW MILLENNIUM

Robin Thomas also presented these reformulations of the four-color theorem at a splendid meeting of the London Mathematical Society on the 150th anniversary of De Morgan's original letter to Hamilton (see Chapter 2). After I gave a survey of the early history of the problem, Appel and Haken (who had flown especially to London) presented an account of some of the main ideas in their solution. Dan Archdeacon of the University of Vermont then explained the Heawood conjecture and other results relating to the coloring of maps on surfaces other than the sphere.

Since the anniversary meeting, good progress has been made in a number of directions. We mention two of these here.

D-REDUCIBLE CONFIGURATIONS

We recall from Chapter 8 that a configuration is *D-reducible* if every coloring of the surrounding ring of countries can be extended to the countries of the configuration and that it is *C*-reducible if it can be proved to be reducible after it has been modified in some way. Since *D*-reducible configurations are much easier to deal with, we may ask (as Walter Stromquist had done in his 1975 thesis) whether there is a proof of the four-color theorem that involves *only* *D*-reducible configurations.

This question was answered in the affirmative by John P. Steinberger in 2008. Inevitably, it involved more configurations and more discharging algorithms than had earlier proofs, and the author found that he needed to consider configurations with larger ring size. The following table summarizes the differences:

	Number of Configurations	Number of Discharging Rules	Largest Ring Size
Appel and Haken	1482	487	14
Robertson et al.	633	32	14
Steinberger	2832	42	16

CHECKING THE PROOF

As we saw in Chapter 11, the correctness of a large computer-assisted proof is difficult to verify. A major breakthrough occurred in September 2004 when the French computer scientist Georges Gonthier provided a fully machine-checked proof of the four-color theorem; this was a formal language implementation and machine verification of the approach of Robertson and his team.

Gonthier's first step was to provide an axiomatic formulation of the terms *map* and *four-colorable*. A respected and widely used machine "proof-checker" called *Coq* then verified some sixty thousand lines of formal language proof before pronouncing the proof valid. At last the doubters could consider the four-color problem proved!

THE FUTURE

Once the four-color problem had been solved, what else was there for map colorers to do? Bill Tutte asked this question back in 1978:

I imagine one of them outgribing in despair, crying 'What shall I do now?' To which the proper answer is 'Be of good cheer. You can continue in the same general line of research.'

For, as we have just seen, the four-color theorem is by no means the end of the line—in fact, it was more of a beginning, as there are several mathematical problems that extend the four-color theorem and develop its ideas in new and exciting directions. Moreover, in spite of its enormous difficulty, proving the four-color problem is just one special instance of some much harder problems, and on these problems good progress has already been made.

With these thoughts of the future in our minds, we leave our last poetic musings to Bill Tutte:

> The Four Colour Theorem is the tip of the iceberg,
> the thin end of the wedge
> and the first cuckoo of Spring.

1750 On November 14 Leonhard Euler states the polyhedron formula in a letter to Christian Goldbach but is unable to prove it.

1794 Adrien-Marie Legendre gives the first correct proof of the polyhedron formula.

1811–13 Simon-Antoine-Jean Lhuilier obtains a version of the polyhedron formula for a polyhedron with tunnels, and Augustin-Louis Cauchy proves the formula for projections of polyhedra onto the plane.

c.1840 August Ferdinand Möbius poses the problem of the five princes to his class.

1852 Francis Guthrie finds that four colors suffice to color a map of England.

On October 23 Augustus De Morgan writes about the four-color problem to Sir William Rowan Hamilton.

1853–54 De Morgan writes about the problem to William Whewell and Robert Ellis.

1854 On June 10 the four-color problem appears in print for the first time, in the *Miscellanea* section of the *Athenaeum*.

1855 Thomas Penyngton Kirkman investigates cycles on polyhedra.

1856 Sir William Rowan Hamilton outlines his icosian calculus and relates it to cycles on a dodecahedron.

1860 On April 14 the four-color problem appears in a book review by De Morgan in the *Athenaeum*.

c.1868 Charles Sanders Peirce presents a solution of the four-color problem at Harvard University.

1878 On June 13 Arthur Cayley asks about the four-color problem at a meeting of the London Mathematical Society.

1879 Arthur Cayley writes a note showing that when trying to solve the four-color problem it is sufficient to consider cubic maps.

Alfred Bray Kempe publishes his purported proof of the four-color theorem in the *American Journal of Mathematics*; Kempe's proof is discussed in scientific gatherings at Johns Hopkins University, Baltimore.

1880 Peter Guthrie Tait shows that the coloring of maps with four colors is equivalent to the coloring of boundary lines with three colors and conjectures that every cubic polyhedron has a Hamiltonian cycle.

Frederick Guthrie identifies his brother Francis as the originator of the four-color problem.

1885 In Leipzig Richard Baltzer confuses the four-color problem with the problem of the five princes.

1886 The four-color problem is posed as a challenge problem for the pupils of Clifton College.

1889 Frederick Temple, Bishop of London, confuses the four-color problem with the problem of the five princes.

1890 Percy Heawood points out the error in Kempe's proof, proves the five-color theorem, and tries to extend the four-color problem to empires and to maps on toruses. He proves that every map on the torus can be colored with seven colors, and produces a torus map that needs seven colors. For a torus with two or more holes, he obtains the correct formula for the number of colors, but fails to prove that there are maps that need this number of colors.

1891 Lothar Heffter points out the deficiency in Heawood's argument for a torus with two or more holes.

1898 Heawood produces a second paper in which he extends Tait's ideas on coloring the boundary lines of a map.

1904 Paul Wernicke produces an unavoidable set of configurations.

1912 George Birkhoff introduces the idea of a chromatic polynomial.

1913 Birkhoff pioneers the study of reducible configurations, proving in particular that the 'Birkhoff diamond' is reducible.

1920 Philip Franklin obtains some new unavoidable sets of configurations and proves that four colors suffice for all maps with up to twenty-five countries.

1926 Clarence Reynolds extends Franklin's result to maps with up to twenty-seven countries.

1930–32 George Birkhoff and Hassler Whitney obtain further results on chromatic polynomials.

c.1935 Heinrich Heesch becomes interested in the four-color problem.

1938 Franklin proves that four colors suffice for all maps with up to thirty-one countries.

1940 C. E. Winn extends Franklin's result to all maps with up to thirty-five countries.

Henri Lebesgue obtains some new unavoidable sets of configurations.

1946 George Birkhoff and D. C. Lewis write a lengthy paper on chromatic polynomials.

Bill Tutte produces a cubic polyhedron that has no Hamiltonian cycle, thereby disproving Tait's conjecture of 1880.

c.1948 Heinrich Heesch proposes to search for an unavoidable set of reducible configurations.

1960s Edward F. Moore shows that any unavoidable set of reducible configurations must necessarily be complicated, containing in particular at least one configuration of ring size 12 or greater.

c.1965 Heinrich Heesch and Karl Dürre use a computer to test the reducibility of configurations.

1967 Oystein Ore publishes the first book on the four-color problem.

1968 Oystein Ore and Joel Stemple prove that four colors suffice for every map with up to forty countries.

Gerhard Ringel and Ted Youngs complete the proof of the Heawood conjecture on the coloring of maps on a torus with at least two holes.

1969 Heinrich Heesch publishes his book on the four-color problem, which includes the first discussion of the method of discharging and introduces the terms *D*-reducible and *C*-reducible.

c.1970 Heinrich Heesch and Wolfgang Haken collaborate on the four-color problem.

1971 Heesch produces his three obstacles to reducibility.

Yoshio Shimamoto produces his horseshoe configuration, which subsequently turns out not to be *D*-reducible.

1972 Kenneth Appel begins his collaboration with Wolfgang Haken.

1974 John Koch joins Appel and Haken.

1975 Martin Gardner produces his April Fool's column in *Scientific American* on a purported counterexample to the four-color problem.

1976 On 22 July Kenneth Appel and Wolfgang Haken publicly announce their proof of the four-color theorem, based on the construction of an unavoidable set of 1936 reducible configurations.

1977 Kenneth Appel, Wolfgang Haken, and John Koch publish their proof of the four-color theorem in the *Illinois Journal of Mathematics*, based on the construction of an unavoidable set of 1482 reducible configurations.

T. L. Saaty and Paul Kainen publish their book on the four-color problem.

1979 Thomas Tymoczko publishes a philosophical paper criticizing Appel and Haken's proof of the four-color theorem.

1981 Ulrich Schmidt discovers an error in the Appel-Haken proof, which is subsequently corrected.

1984 Brad Jackson and Gerhard Ringel solve Heawood's empire problem in general.

1986 Appel and Haken publish a paper describing their methods and refuting the persistent rumours about their proof.

1989 Appel and Haken produce an extended version of their proof in book form: *Every Planar Map is Four Colorable*.

1994 Neil Robertson, Daniel Sanders, Paul Seymour, and Robin Thomas obtain a revised proof of the four-color theorem. Following Appel and Haken's general approach, they produce an unavoidable set with 633 reducible configurations, using a computer for both parts of the proof.

2002 The 150th anniversary of the first appearance of the four-color problem is celebrated at a special meeting of the London Mathematical Society.

2004 Georges Gonthier provides a formal machine verification of the Robertson et al. proof.

2008 John P. Steinberger produces a proof of the four-color theorem that involves only *D*-reducible configurations.

FURTHER READING

For an overview of the history of the four-color problem up to 1936, including reprints of several of the papers described here, see the 1998 paperback edition of N. L. Biggs, E. K. Lloyd, and R. J. Wilson, *Graph Theory 1736–1936*, Clarendon Press, Oxford. This book is referred to here as *BLW*.

A detailed account of the underlying mathematics in the language of graph theory is given in Rudolf Fritsch and Gerda Fritsch, *The Four-Color Theorem: History, Topological Foundations, and Ideas of Proof*, Springer, 1999; Chapter 1 contains biographies of several of the mathematicians featured here. This book is referred to below as *FF*.

A well-written textbook on graph theory, concentrating particularly on the four-color problem for graphs, is by my namesake, Robert A. Wilson, *Graphs, Colourings and the Four-Colour Theorem*, Oxford Science Publications, 2002.

Earlier influential books on the four-color problem are O. Ore, *The Four-Color Problem*, Academic Press, 1967, and Thomas L. Saaty and Paul C. Kainen, *The Four-Color Problem: Assaults and Conquest*, McGraw-Hill, 1977 (reprinted in paperback by Dover, 1986).

For Kenneth Appel and Wolfgang Haken's own accounts of the history of the four-color problem and of their part in its solution, see "The solution of the four-color-map problem", *Scientific American* **237** No. 4 (October 1977), 108–21, and "The four-color problem", in *Mathematics Today* (ed. L. A. Steen), Springer (1978), 153–80.

There is also an excellent account of the history and proof of the four-color problem, especially on the work of Heesch, Haken, and Appel, and the philosophical aspects of the solution, in Donald MacKenzie, "Slaying the Kraken: the sociohistory of a mathematical proof", *Social Studies of Science* **29** (1) (February 1999), 7–60.

Another invaluable source for the writing of this book was the unpublished manuscript of a proposed book, *Origins of Graph Theory*, by P. J. Federico, who unfortunately died before he was able to bring his project to completion.

Biographies of several of the mathematicians mentioned in this book appear in the *Dictionary of Scientific Biography* (ed. C. C. Gillespie), Scribner's, New York, 1970–1990 (referred to below as *DSB*), and in the four-volume version, *Biographical Dictionary of Mathematicians* (1991), extracted from the *DSB*. Among the many general histories of mathematics that can be recommended are Dirk J. Struik, *A Concise History of Mathematics* (4th ed.), Dover Publications, 1967, and Victor J. Katz, *A History of Mathematics: An Introduction* (3d ed.), Pearson, 2008.

DETAILED NOTES AND REFERENCES

1 The Four-Color Problem

2. Kenneth O. May's article on the four-color problem is "The origin of the four-color conjecture", *Isis* **56** (1965), 346–8. Another article on the problem's origins is H.S.M. Coxeter's "The four-color map problem, 1840–1940", *Mathematics Teacher* **52** (1959), 283–9.

2. The book that asserts that graph theory developed primarily from work on the four-color theorem is M. Aigner, *Graphentheorie—Eine Entwick-lung aus dem 4-Farbenproblem*, B. G. Teubner, Stuttgart, 1984.

10. The April Fool's column and the follow-up appeared in Martin Gardner, "Six sensational discoveries that somehow have escaped public attention", *Scientific American* **232** No. 4 (April 1975), 126–30, and **233** No. 1 (July 1975), 115.

2 The Problem Is Posed

Biographies of Augustus De Morgan, Sir William Rowan Hamilton, Charles Sanders Peirce, and August Ferdinand Möbius appear in the *DSB*, and and biographies of Francis Guthrie appear in the *Dictionary of South African Biography* **2** (1972), 279–80, and by Pieter Maritz and Sonja Mouton, "Francis Guthrie: a colourful life" in *The Mathematical Intelligencer* **34**(3) (2012), 67–75. Brief biographies of Francis and Frederick Guthrie, De Morgan, Hamilton, Whewell, Ellis, and Peirce appear in Chapter 1 of *FF*.

12. De Morgan's letter to Sir William Rowan Hamilton, sent from 7 Camden Street, Camden Town, on October 23, 1852, is now housed in the archive collection of the library of Trinity College, Dublin (TCD MS 1493/668).

14. Frederick Guthrie's "Note on the colouring of maps" appeared in the *Proceedings of the Royal Society of Edinburgh* **10** (1880), 727–8.

15. Heinrich Tietze's example appeared in his *Famous Problems of Mathematics*, Graylock Press, New York (1965), 77–8.

16. The meeting between De Morgan and Hamilton is recalled by De Morgan's widow Sophia in S. E. De Morgan, *Memoir of Augustus De Morgan, with Selections of His Writings*, Longman, Green & Co., London (1882), 333.

17. Hamilton's reply to De Morgan was sent from Dunsink Observatory, Dublin, on October 26, 1852.

De Morgan's letters to William Whewell and Robert Ellis, dated December 9, 1853, and June 24, 1854, are housed in the Wren Library, Trinity College, Cambridge (Whewell Add. Mss., a.202^{125} and c.67^{111}).

A useful discussion of De Morgan's ideas appears in N. L. Biggs, "De Morgan on map colouring and the separation axiom", *Archive for History of Exact Sciences* **28** (1983), 165–70.

18. De Morgan's anonymous review of W. Whewell's *The Philosophy of Discovery* appears in *The Athenaeum*, No. 1694 (April 14, 1860), 501–3. The later rediscovery of this review was by John Wilson, "New light on the origin of the four-colour conjecture", *Historia Mathematica* **3** (1976), 329.

The Hotspur quotation appears in William Shakespeare's *King Henry IV, Part 1*, Act III, Scene I.

19. De Morgan's March 3, 1860. letter to Whewell appears on p. 302 of S. E. De Morgan's book (see p. 16). The earlier *Athenaeum* reference was found by Brendan D. McKay and is described in his article "A note on the history of the four-colour conjecture", *Journal of Graph Theory* **72** (2013), 361–63.

Peirce's interest in the four-color problem is described in Chapter 19, "The four-color problem", in *Studies in the Scientific and Mathematical Philosophy of Charles S. Peirce: Essays by Carolyn Eisele* (ed. R. M. Martin), Mouton, The Hague, 1971, 216–22, and is referred to in Norman L. Biggs, E. Keith Lloyd and Robin J. Wilson, "C. S. Peirce and De Morgan on the four-colour conjecture", *Historia Mathematica* **4** (1977), 215–16.

The Peirce quotation appears on p. 219 of Eisele (see above).

21. The discovery of Listing and Möbius's one-sided surface is outlined in P. Stäckel, "Die Entdeckung der einseitigen Flächen", *Mathematische Annalen*

22. The origin of Möbius's problem of the five princes is presented in R. Baltzer, "Eine Erinnerung an Möbius und seinen Freund Weiske", *Berichte der Sächsischen Gesellschaft der Wissenschaften zu Leipzig* **37** (1885), 1–6.

The problem of the five palaces is discussed in Tietze's book (see p. 15).

26. The content of Baltzer's lecture is summarized in his article (see p. 22).

27. Baltzer's error was compounded by F. Dingeldey in *Topologische Studien*, B. G. Teubner, Leipzig, 1890; by Isabel Maddison, "Note on the history of the map-coloring problem", *Bulletin of the American Mathematical Society* **3** (1896–7), 257; by W. W. Rouse Ball, *Mathematical Recreations and Essays* (11th ed.), Macmillan, London (1939), p. 223; and by E. T. Bell, *The Development of Mathematics* (2nd ed.), McGraw-Hill (1945), p. 606. The record on Möbius's contribution to the four-color problem was set straight by H.S.M. Coxeter (see p. 2).

3 Euler's Famous Formula

Biographies of Leonhard Euler, Augustin-Louis Cauchy, and Simon-Antoine-Jean Lhuilier appear in the *DSB*. For a readable account of Euler's life see Emil A. Fellmann, *Leonhard Euler*, Birkhäuser-Verlag, 2007, and for a more mathematical treatment see William Dunham's *Euler: The Master of Us All*, Mathematical Association of America, 1999. For a fuller treatment of Euler's polyhedron formula and a discussion of several of its consequences, see David S. Richeson's *Euler's Gem: The Polyhedron Formula and the Birth of Topology*, Princeton University Press, 2008.

30. For an excellent account of the history and properties of polyhedra, see Peter R. Cromwell, *Polyhedra*, Cambridge University Press, 1997 (paperback ed., 1999).

For a geometrical proof that there are only five regular polyhedra, see Book XIII in T. L. Heath's *The Thirteen Books of Euclid's Elements*, Cambridge University Press, 1908. Further details can be found in Cromwell's book (see above).

33. The extract from Euler's letter to Christian Goldbach appears in Leonhard Euler's *Opera Omnia* (4), I, letter 863. It also appears in printed form in P.-H. Fuss, *Correspondance Mathématique et Physique de Quelques Célèbres Géomètres*

du XVIIIème Siècle, St. Petersburg, 1843; an English translation appears in *BLW*, pp. 76–7. The letter appears in full in A. P. Juškević and E. Winter, *Leonhard Euler und Christian Goldbach: Briefwechsel 1729–1764*, Akademie-Verlag, Berlin, 1965.

34. For accounts of René Descartes's work on polyhedra, see Cromwell's book (see p. 41), C. Adam and P. Tannery, *Oeuvres de Descartes*, Vol. 10, Cerf, Paris (1897–1913), 257–77; and P. J. Federico's *Descartes on Polyhedra: A Study of the De Solidorum Elementis*, Springer-Verlag, 1982.

35. Euler's papers on polyhedra are "Elementa doctrinae solidorum" and "Demonstratio nonnullarum insignium proprietatum quibus solida hedris planis inclusa sunt praedita", *Novi Commentarii Academiae Scientiarum Imperialis Petropolitanae* **4** (1752–3, publ. 1758), 109–40, 140–60; both papers appear in Vol. 26 of *Leonhardi Euleri Opera Omnia (1)* (ed. A. Speiser), Commentationes Geometricae, Zürich, 1953.

    Legendre's proof appeared in his *Eléments de Géométrie* (1st ed.), Firmin Didot, Paris, 1794.

36. Cauchy's paper on polyhedra was "Recherches sur les polyèdres—premier mémoire", *Journal de l'Ecole Polytechnique* **9** (Cah. 16) (1813), 68–86; an English translation of part of this paper appears in *BLW*, pp. 81–3.

38. The examples given by Lhuilier (whose name is also written as "L'Huilier") appeared in "Mémoire sur la polyédrométrie", *Annales de Mathématiques* **3** (1812–13), 169–89; an English translation of part of this paper appears in *BLW*, pp. 84–6.

39. The "only five neighbors theorem" for maps was obtained by Kempe (see Chapter 5), who also derived a result similar to the counting formula.

4 Cayley Revives the Problem . . .

Brief biographies of Arthur Cayley appear in the DSB and Chapter 1 of FF. A more substantial work is Tony Crilly's major biography *Arthur Cayley: Mathematician Laureate of the Victorian Age*, Johns Hopkins University Press, 2006.

46. Cayley's query appeared in the *Proceedings of the London Mathematical Society* **9** (1877–8), 148; and in *Nature* **18** (1878), 294. His paper "On the colouring of maps" appears in the *Proceedings of the Royal Geographical Society* **1** (1879), 259–61; it is reprinted in full in *BLW*, pp. 93–4.

47. For the full story of Galton's involvement with the four-color problem, see Tony Crilly, "Arthur Cayley FRS and the four-colour map problem", *Notes & Records of the Royal Society* **59** (2005), 285–304; Cayley drafted two versions of his paper for the *Proceedings of the Royal Geographical Society*, both of which are carefully analyzed in Crilly's paper.

49. Levi ben Gerson's use of mathematical induction is discussed in Victor Katz, "Combinatorics and induction in medieval Hebrew and Islamic mathematics", *Vita Mathematica* (ed. R. Calinger), Mathematical Association of America (1996), 99–106.

5 . . . and Kempe Solves It

A brief biography of Sylvester appears in the DSB. More substantial works are Karen Hunger Parshall's major biography *James Joseph Sylvester: Jewish*

Mathematician in a Victorian World, Johns Hopkins University Press, 2006, and her book on his letters (see p. 75). A biography of Alfred Bray Kempe (by Archibald Geikie) appeared in the *Proceedings of the Royal Society* **102** (1923), i–x. Brief biographies of Sylvester, Kempe, and William Story are given in Chapter 1 of *FF*.

56. Sylvester's correspondence with Gilman and others appears in Karen Parshall, *James Joseph Sylvester: Life and Work in Letters*, Clarendon Press, Oxford, 1998.

57. Kempe's first paper was "On the solution of equations by mechanical means", *Messenger of Mathematics* **2** (1873), 51–2. His "How to draw a straight line" was first published in four parts in *Nature* **16** (1877), 65–7, 86–9, 125–7, and 145–6.

58. Kempe's paper "On the geographical problem of the four colours" appeared in the *American Journal of Mathematics* **2** (part 3) (1879), 193–200; it is reprinted in full in *BLW*, pp. 96–102. A preview of it appeared in *Nature* **20** (July 17, 1879), 275; and simplified versions appeared in the *Proceedings of the London Mathematical Society* **10** (1878–9), 229–31, and in *Nature* **21** (February 26, 1880), 399–400.

67. The term *graph* was first used in this sense by Sylvester in "Chemistry and algebra", *Nature* **17** (1877–8), 284; this note is reprinted in full in *BLW*, pp. 65–6.

68. Story's "Note on the preceding paper" appeared in the *American Journal of Mathematics* **2** (1879), 201–4.

Sylvester's letter to D. Gilman appears on pp. 195–6 of Parshall's book (see p. 56).

Reports of the November 5 and December 3, 1879, meetings of the Johns Hopkins Scientific Association appeared in the *Johns Hopkins University Circular* **1**, No. 2 (January 1880), 16.

69. Peirce's reformulation of the four-color problem appears on p. 218 of Eisele (see p. 26). The "Notice by Mr. Peirce" appeared in *The Nation* No. 756 (December 25, 1879), 440.

Peirce's New York lecture was mentioned in the *Report of the National Academy of Sciences for the Year 1899* (1900), 12–13.

70. The letter from Story to Peirce appears on p. 359 of Eisele (see p. 26).

6 A Chapter of Accidents

Biographies of Charles Dodgson, Peter Guthrie Tait, Thomas Penyngton Kirkman, and Sir William Rowan Hamilton appear in the *DSB*. A popular account of Dodgson's mathematical writings can be found in Robin Wilson's *Lewis Carroll in Numberland: His Fantastical Mathematical Logical Life*, Allen Lane, 2008. For full biographies of Tait and Hamilton, see C. C. Knott, *Life and Scientific Work of Peter Guthrie Tait*, Cambridge University Press, 1911, and R. P. Graves, *Life of Sir William Rowan Hamilton*, Longman, London, and Hodges, Figgis & Co., Dublin, 1889. Kirkman's life and works are chronicled in N. L. Biggs, "T. P. Kirkman, mathematician", *Bulletin of the London Mathematical Society* **13** (1981), 97–120. Brief biographies of Tait and Hamilton appear in Chapter 1 of *FF*.

71. Dodgson's *An Elementary Theory of Determinants* was published by Macmillan

in 1867. His puzzle appeared in Stuart D. Collingwood's *The Life and Letters of Lewis Carroll (Rev. C. L. Dodgson)*, Nelson, 1898, 371.

Edouard Lucas's translation of Kempe's paper, "Le Problème géographique des quatre coulours', appeared in *Revue Scientifique (3)* **6** (1883), 12–17.

72. Lucas's extended discussion of the four-color problem appeared in *Récréations Mathématiques*, Vol. 4, Gauthier-Villars, 1894.

For information about Baltzer's lecture to the Leipzig Scientific Society, see p. 20.

73. J. M. Wilson's letters of January 1, 1887, and June 1, 1889, appeared in the *Journal of Education* **9** (1887), 11–12, and **11** (1889), 277. His account of Frederick Temple's solution also appears in *Memoirs of Archbishop Temple*, Vol. 1 (ed. E. G. Sandford), Macmillan, London, 1906, 59–60.

73. John Cook Wilson's paper "On a supposed solution of the 'four-colour problem'" appeared in the *Mathematical Gazette* **3** (1904–6), 338–40.

74. The golf ball anecdote is discussed by Chris Denley and Chris Pritchard in "The golf ball aerodynamics of Peter Guthrie Tait", *Mathematical Gazette* **78** (1994), 298–313.

74. For Kempe's note in *Nature*, see p. 58.

P. G. Tait's first attempted solutions appeared in "On the colouring of maps", *Proceedings of the Royal Society of Edinburgh* **10** (1878–80), 501–3.

77. Tait's second note on the subject was "Remarks on the previous communication" [by Guthrie], *Proceedings of the Royal Society of Edinburgh* **10** (1878–80), 729; it appears in full in *BLW*, p. 104. His letters to Kempe are in the West Sussex Public Record Office, Chichester.

For details of Frederick Guthrie's note, see p. 46.

78. Tait's paper was "Note on a theorem in the geometry of position", *Transactions of the Royal Society of Edinburgh* **29** (1880), 657–60.

80. Kirkman's paper "On the representation of polyhedra" appeared in the *Philosophical Transactions of the Royal Society* **146** (1856), 413–18; part of this paper appears in *BLW*, pp. 29–30.

81. W. R. Hamilton's icosian calculus was described in "Memorandum representing a new system of roots of unity", *Philosophical Magazine (4)* **12** (1856), 446, and in the *Proceedings of the Royal Irish Academy* **6** (1853–7), 415–16. A summary of the main ideas, together with instructions for playing the icosian game, can be found in *BLW*, pp. 32–5.

82. For details of Tait's paper, see p. 78.

Tait's final paper on the subject was "Listing's *Topologie*", *Philosophical Magazine (5)* **17** (1884), 30–46.

84. Kirkman's poem and discussion appeared as Question 6610 with Solution by the proposer in *Mathematical Questions and Solutions from the Educational Times* **35** (1881), 112–16.

85. W. T. Tutte's cubic polyhedron appeared in "On Hamiltonian circuits", *Journal of the London Mathematical Society* **21** (1946), 98–101.

The poem about Bill Tutte is by Norman Biggs.

7 A Bombshell from Durham

A biography of Julius Petersen appears in the *DSB*. A full account of Heawood's life is G. A. Dirac's "Percy John Heawood". *Journal of the London Mathematical*

Society **38** (1963), 263–77; a brief obituary by J. Duff appeared in *Nature* **175** (1955), 368, and a brief biography appears in Chapter 1 of *FF*.

88. Heawood's letter to Alfred Errera appeared in Errera's "Exposé historique du problème des quatre couleurs", *Periodica Mathematica (4)* **7** (1927), 20–41.

89. Heawood's "Map-colour theorem" appeared in the *Quarterly Journal of Pure and Applied Mathematics* **24** (1890), 332–8; most of it can be found in *BLW*, pp. 105–7, 112–15.

91. Kempe's admission of his error is reported in the *Proceedings of the London Mathematical Society* **22** (1890–91), 263.

92. Errera's example appeared in his doctoral thesis, published as *Du Coloriage des Cartes et de Quelques Questions d'Analysis Situs*, Falk Fils, Brussels and Gauthier-Villars, Paris, 1921; it is discussed by Joan Hutchinson and Stan Wagon in "Kempe revisited", *American Mathematical Monthly* **105** (February 1998), 170–74. De la Vallée Poussin's example appeared in "Deuxième réponse à Question 51", *L'Intermédiaire des Mathématiciens* **3** (1896), 179–80.

96. For the general solution of the empire problem, see Brad Jackson and Gerhard Ringel, "Heawood's empire problem", *Journal of Combinatorial Theory B* **38** (1985), 168–78.

97. A map that needs nine colors for the earth-moon problem was found by Thom Sulanke in 1973; his map appears in Martin Gardner's "Mathematical Games" column, *Scientific American* 242(2) (February 1980), 14–19.

101. L. Heffter's paper is "Ueber das Problem der Nachbargebiete", *Mathematische Annalen* **38** (1891), 477–508; an English translation of part of it appears in *BLW*, pp. 118–23.

102. For Lucas's *Récréations Mathématiques*, see p. 72.

   The Paris papers were by P. Mansion, "Question 51"; H. Delannoy and A. S. Ramsey, "Réponse à Question 51"; C. de la Vallée Poussin, "Deuxième réponse"; and H. Delannoy, "Troisième réponse", in *L'Intermédiaire des Mathématiciens* **1** (1894), 20, 192 and **3** (1896), 179–80, 225.

   Julius Petersen's two notes appeared in *L'Intermédiaire des Mathématiciens* **5** (1998), 225–7 and **6** (1899), 36–8.

103. P. J. Heawood's second paper, "On the four-colour map theorem", appeared in the *Quarterly Journal of Pure and Applied Mathematics* **29** (1898), 270–85.

104. Heawood's probability arguments appeared in "Failures in congruences connected with the four-colour map theorem", *Proceedings of the London Mathematical Society (2)* **40** (1936), 189–202.

8 Crossing the Atlantic

Biographies of George David Birkhoff and Oswald Veblen appear in the *DSB*. A biography of Birkhoff appears in Marston Morse, "George David Birkhoff and his mathematical work", *Bulletin of the American Mathematical Society* **52** (1946), 357–91, and a brief biography appears in Chapter 1 of *FF*.

105. The Minkowski story appears in Constance Reid's *Hilbert*, Springer (1970), 92–3.

107. The information about Wernicke appears in P. J. Federico's manuscript (see Further Reading, above).

108. For a brief report of Wernicke's Toronto lecture, see the *Bulletin of the American Mathematical Society* **4** (1897–8), 2, 5.

Wernicke's main paper is "Über den kartographischen Vierfarbensatz", *Mathematische Annalen* **58** (1904), 413–26.

Heesch's "method of discharging" appeared in his *Untersuchungen zum Vierfarbenproblem*, B. I. Hochschulskripten, 810/810a/ 810b, Bibliographisches Institut, Mannheim-Wien-Zürich, 1969.

111. Many of the results in Philip Franklin's doctoral thesis, "On the map color problem", Princeton University, 1921, appeared in his paper "The four color problem", *American Journal of Mathematics* **44** (3) (July 1922), 225–36; this paper appears in full in *BLW*, pp. 171–80, and he later wrote two introductory articles, "The four color problem", *Scripta Mathematica* **6** (1939), 149–56, 197–210. A useful survey paper from the 1920s is H. R. Brahana, "The four-color problem", *American Mathematical Monthly* **30** (July/August 1923), 234–43.

Henri Lebesgue's paper was "Quelques conséquences simple de la formule d'Euler", *Journal de Mathématiques Pures et Appliquées* **9** (1940), 27–43.

112. Veblen's paper was "An application of modular equations in analysis situs", *Annals of Mathematics (2)* **14** (1912–3), 86–94; it appears in full in *BLW*, pp. 160–66.

Birkhoff's pioneering paper was "The reducibility of maps", *American Journal of Mathematics* **35** (1913), 115–28.

115. Arthur Bernhart's paper was "Six-rings in minimal five-color maps", *American Journal of Mathematics* **69** (1947), 391–412. The honeymoon anecdote is from a personal communication by Garrett Birkhoff.

The remark about the Kohinoor diamond appears on page 156 of *FF*. For the theses of Franklin and Errera, see pp. 111 and 92.

116. For the contributions of Reynolds, Franklin, and Winn, see C. N. Reynolds, "On the problem of coloring maps in four colors I, II", *Annals of Mathematics (2)* **28** (1926–7), 1–15, 477–92; P. Franklin, "Note on the four color problem," *Journal of Mathematics and Physics* **16** (1938), 172–84; and C. E. Winn, "On the minimum number of polygons in an irreducible map", *American Journal of Mathematics* **62** (1940), 406–16. The work of Paul Valéry is mentioned on p. 32 of *FF* and is described in J. Mayer, "Une page mathématique de Valéry: le problème du coloriage des cartes", *Bulletin Etudes Valéryennes* **25** (1980), 31–43.

119. For Heesch's classification of configurations, see his book cited on p. 108.

120. Birkhoff's first paper on chromatic polynomials was "A determinant formula for the number of ways of coloring a map", *Annals of Mathematics (2)* **14** (1912–13), 42–6; part of this paper appears in *BLW*, pp. 167–9. Birkhoff's paper with D. C. Lewis is "Chromatic polynomials", *Transactions of the American Mathematical Society* **60** (1946), 355–451.

121. Hassler Whitney's contribution to chromatic polynomials can be found in "A logical expansion in mathematics", *Bulletin of the American Mathematical Society* **38** (1932), 572–9; part of this paper appears in *BLW*, pp. 181–4. Whitney went on to write some fundamental papers in the related area of graph theory, before turning his attention to the field of algebraic topology, in which he became one of the leading exponents of his generation.

122. The connections between chromatic polynomials and the golden ratio are outlined in G. Berman and W. T. Tutte, "The golden root of a chromatic

polynomial", *Journal of Combinatorial Theory* **6** (1969), 301–2; and explored in W. T. Tutte, "On chromatic polynomials and the golden ratio", *Journal of Combinatorial Theory* **9** (1970), 289–96.

9 A New Dawn Breaks

For a biography of Heinrich Heesch, see Hans-Günther Bigalke, *Heinrich Heesch: Kristallgeometrie, Parkettierungen, Vierfarbenforschung*, Birkhäuser, Basel, 1988. Brief biographies of Heesch, Haken, and Dürre appear in Chapter 1 of *FF*. The quotations by Wolfgang Haken in this chapter are taken from an interview conducted by Tony Dale on April 16, 1994, in Urbana, Illinois, for Donald MacKenzie's "Slaying the Kraken" (see Further Reading, above).

For the papers by Wernicke, Franklin, and Lebesgue, see pp. 107 and 111.

124. Dénes König's book is *Theorie der Endlichen und Unendlichen Graphen*, Akademische Verlagsgesellschaft, Leipzig, 1936; an English edition, *Theory of Finite and Infinite Graphs*, was published by Birkhäuser in 1990.

The graph theory texts by Berge, Ore, Busacker and Saaty, and Harary are: C. Berge, *Theory of Graphs and Its Applications*, Wiley, 1961; O. Ore, *Theory of Graphs*, American Mathematical Society, 1961; R. G. Busacker and T. L. Saaty, *Finite Graphs and Networks*, McGraw-Hill, 1965; and F. Harary, *Graph Theory*, Addison-Wesley, 1969.

For details of Oystein Ore's book *The Four-Color Problem*, see Further Reading, above. The paper by O. Ore and J. Stemple is "Numerical calculations on the four-color problem", *Journal of Combinatorial Theory* **8** (1970), 65–78.

125. A full discussion of the solution of the Heawood conjecture appears in G. Ringel, *Map Color Theorem*, Springer, 1974. Further information can be found in G. Ringel and J.W.T. Youngs, "Solution of the Heawood map-coloring problem", *Proceedings of the National Academy of Sciences, U.S.A.* **60** (1968), 438–45.

127. David Hilbert's lecture, "Sur les problèmes futurs des mathématiques", appeared in the *Proceedings of the Second International Congress of Mathematicians, Paris* (1902), 58–114. There is a good account of his problems in J. J. Gray, *The Hilbert Challenge*, Oxford University Press, 2000. Heesch's work on tilings is discussed in detail in Bigalke's book, cited above.

129. Wolfgang Haken's Amsterdam lecture on the knot problem is mentioned briefly in the *Proceedings of the International Congress of Mathematicians*, Amsterdam, 1954. A readable account of Perelman's solution of the Poincaré conjecture can be found in Donal O'Shea's book *The Poincaré Conjecture: In Search of the Shape of the Universe*, Allen Lane, 2007, and his solution appears in "Theorie der Normalflächen", *Acta Mathematica* **105** (1961), 245–375.

130. For Kurt Gödel's 1931 paper in English, "On formally undecidable propositions of the *Principia Mathematica*", see *From Frege to Gödel* (ed. J. van Heijenoort), Harvard University Press, 1967, 596–616.

131. For Birkhoff's paper, see p. 112, above; for Heesch's classification of configurations, see p. 108.

133. Edward F. Moore's map appears in L. Steen, "Solution of the four color problem", *Mathematics Magazine* **49** (4) (September 1976), 219–22; the map that appears in *Scientific American* **237** No. 4 (October 1977), 109, is incorrect.

134. For details of Oystein Ore's book on the four-color problem, see Further Reading, above.

138. Yoshio Shimamoto's remark appears on p. 212 of Bigalke's book *Heinrich Heesch* (see above).

The paper by Hassler Whitney and Bill Tutte is "Kempe chains and the four-colour problem", *Utilitas Mathematicae* **2** (1972), 241–81; it was reprinted in *Studies in Graph Theory II* (ed. D. R. Fulkerson), Mathematical Association of America (1975), 378–413.

10 Success!

Brief biographies of Wolfgang Haken, Kenneth Appel, and John Koch appear in Chapter 1 of *FF*. For a biography of Heinrich Heesch, see Hans-Günther Bigalke, *Heinrich Heesch: Kristallgeometrie, Parkettierungen, Vierfarbenforschung*, Birkhäuser, Basel, 1988. The quotations by Haken, Appel, and Koch in this chapter are taken from interviews conducted by Tony Dale on April 16, June 6, and May 3, 1994, for Donald MacKenzie's "Slaying the Kraken" (see Further Reading, above).

139. Some of Heesch's 8900 configurations appear on p. 203 of Bigalke's book *Heinrich Heesch* (see above).

140. Walter Stromquist's paper proving the four-color theorem for all maps with up to 51 countries is "The four-color theorem for small maps", *Journal of Combinatorial Theory B* **19** (1975), 256–68. His doctoral thesis was "Some aspects of the four color problem", Harvard University, 1975.

141. Shimamoto's remark appears on p. 224 of Bigalke's book *Heinrich Heesch* (see above).

144. Appel and Haken's paper on "The existence of unavoidable sets of geographically good configurations" appeared in the *Illinois Journal of Mathematics* **20** (1976), 218–97. For their paper on maps with no adjacent pentagons, "An unavoidable set of configurations in planar triangulations", see the *Journal of Combinatorial Theory B* **26** (1979), 1–21.

146. Koch's doctoral thesis was "Computation of four color irreducibility", University of Illinois at Urbana-Champaign, 1976.

150. Allaire and Swart's joint paper, "A systematic approach to the determination of reducible configurations in the four-color conjecture", appeared in the *Journal of Combinatorial Theory B* **25** (1978), 339–62.

151. After the Appel-Haken solution appeared, Frank Bernhart wrote "A digest of the four color theorem", *Journal of Graph Theory* **1** (1977), 207–25.

152. Tutte's earlier paper was "Map coloring problems and chromatic polynomials", *American Scientist* **62** (1974), 702–5.

153. Reports on the Appel-Haken solution appeared in many newspapers and periodicals of 1976, including *The Times* (July 23), *SIAM News* (August), *Science* (August 13), the *Toronto Globe* (August 24), *Le Monde* (September 1), *Time* (September 20), the *New York Times* (September 24 and 26), *Scientific American* (October), *New Scientist* (October 21), and *Die Neue Zürcher Zeitung* (October 24).

154. The official research announcement was K. Appel and W. Haken, "Every planar map is four colorable", *Bulletin of the American Mathematical Society* **82** (1976), 711–12.

156. The solution to the four-color problem appeared in two parts, together with a microfiche: K. Appel and W. Haken, "Every planar map is four colorable, Part I: Discharging", *Illinois Journal of Mathematics* **21** (1977), 429–90; and K. Appel, W. Haken, and J. Koch, "Every planar map is four colorable, Part II: Reducibility", *Illinois Journal of Mathematics* **21** (1977), 491–567.

11 Is It a Proof?

The quotations by Haken, Appel, and Swart in this chapter are taken from interviews conducted by Tony Dale on April 16, June 6, and June 1, 1994, for Donald MacKenzie's "Slaying the Kraken" (see Further Reading, above).

157. The quotation "in my view . . ." appears in F. F. Bonsall, "A down-to-earth view of mathematics", *American Mathematical Monthly* **89** (1982), 8–15; the remark "God wouldn't let the theorem . . ." was made by Herbert Wilf to Kenneth Appel in mid-1976.

Donald Albers' report "Polite applause for a proof of one of the great conjectures of mathematics: what is a proof today?" appeared in the *Two-Year College Mathematics Journal* **12** (2) (March 1981), 82.

158. The experience of Armin Haken is recalled in the previously mentioned interviews with Appel and Haken.

159. Thomas Tymoczko's article, "The four-color problem and its philosophical significance" appeared in *The Journal of Philosophy* **76** (2) (February 1979), 57–83. Another article by him was "Computers, proofs and mathematicians: a philosophical investigation of the four-color proof", *Mathematics Magazine* **53** (3) (May 1980), 131–8.

160. Ted Swart's response was "The philosophical implications of the four-color problem", *American Mathematical Monthly* **87** (November 1980), 697–707.

161. The group theory articles referred to are W. Feit and J. G. Thompson, "Solvability of groups of odd order", *Pacific Journal of Mathematics* **13** (1963), 775–1029; and D. Gorenstein, "The classification of finite simple groups I", *Bulletin of the American Mathematical Society* **4** (1) (January 1979), 43–200. Andrew Wiles's proof is described in Simon Singh's *Fermat's Last Theorem*, Fourth Estate, London, 1997.

162. Ian Stewart's comment appears on p. 304 of *Concepts of Modern Mathematics*, Penguin, 1981. Daniel Cohen's remarks appeared in "The superfluous paradigm", *The Mathematical Revolution Inspired by Computing* (ed. J. H. Johnson and M. J. Loomes), Oxford (1991), 323–9. H.S.M. Coxeter's comments were made in an interview by D. Albers and G. L. Alexanderson published in *Mathematical People*, Birkhäuser, 1985. George Spencer-Brown's comments appear in his solution to the four-color problem, published as an appendix to the German translation of his *Laws of Form*, 1997.

G. H. Hardy's remark appears in Section 10 of *A Mathematician's Apology*, Cambridge University Press, 1940.

163. Accounts of G. Spencer-Brown's solution and his lecture appeared in the *Times Higher Education Supplement*, December 17 and 24, 1976, and in *New Scientist*, December 23, 30, 1976, and January 6, 1977. His visit to Stanford University is mentioned in Martin Gardner's *The Last Recreations*, Springer/Copernicus, 1997, 88–9. Further discussion of Spencer-Brown's ideas, linking them with other areas of mathematics, appears in Louis H. Kauffman,

"Reformulating the map color theorem", *Discrete Mathematics* **202** (2005), 145–72.

164. Frank Allaire's paper "Another proof of the four colour theorem, I" appeared in the *Proceedings of the Seventh Manitoba Conference on Numerical Mathematics and Computing* (1977), 3–72.

Appel and Haken's *Scientific American* article and Saaty and Kainen's book are cited in Further Reading, above. T. L. Saaty's article "Thirteen colorful variations on Guthrie's four-color conjecture" appeared in the *American Mathematical Monthly* **79** (January 1972), 2–43.

165. Ulrich Schmidt's dissertation was "Überprüfung des Beweis für den Vierfarbensatz", Diplomarbeit, Technische Hochschule Aachen, 1982. S. Saeki's observations appear in his "Verification of the Discharging Procedure in the Four Color Theorem", Master's Thesis, University of Tokyo, 1985.

K. Appel and W. Haken's article "The four color proof suffices" appeared in the *Mathematical Intelligencer* **8** (1) (1986), 10–20, 58. Their book *Every Planar Map is Four Colorable* was published by the American Mathematical Society in 1989.

166. The proof by Neil Robertson, Daniel Sanders, Paul Seymour, and Robin Thomas was outlined by Seymour in "Progress on the four-color theorem", *Proceedings of the International Congress of Mathematicians, Zürich*, Birkhäuser, 1995, and was described by all four of them in "A new proof of the four-colour theorem", *Electronic Research Announcements of the American Mathematical Society* **2** (1) (August 1996), 17–25, and in "The four-colour theorem", *Journal of Combinatorial Theory B* **70** (1997), 2–44.

167. Robin Thomas's article was "An update on the four-color theorem", *Notices of the American Mathematical Society* **45** (7) (August 1998), 848–59.

The talks given at the 150th anniversary meeting in London were:

> Robin Wilson: *"The four colour problem: 1852–1940"*
> Kenneth Appel and Wolfgang Haken: *"Solving the four-colour problem"*
> Dan Archdeacon: *"From the Heawood conjecture to topological graph theory"*
> Robin Thomas: *"The four-colour theorem and beyond"*

168. John P. Steinberger's article was "An unavoidable set of D-reducible configurations", *Transactions of the American Mathematical Society* **362** (2010), 6633–61.

An article by Georges Gonthier about his computer verification appears in "Formal proof—The four-color theorem", *Notices of the American Mathematical Society* **55** (December 2008), 1382–93.

W. T. Tutte's article "Colouring problems" appeared in the *Mathematical Intelligencer* **1** (1978), 72–5.

169. Among the "much harder problems" that remain unsolved are the finding of proofs of Hadwiger's conjecture and the five-flow conjecture. These are expressed in the language of graphs and details of them can be found in books on graph theory.

It is very appropriate that the final words of this book are by Bill Tutte. Not only was he one of the greatest graph theorists of the twentieth century, but the sad news of his death reached me on the day I handed the original manuscript of this book to the publisher.

algorithm A finite step-by-step method for solving a problem.

Archimedean polyhedron See **semiregular polyhedron**.

Birkhoff diamond A configuration of four adjoining pentagons surrounded by a ring of six countries.

boundary line Part of the boundary of a country, connecting two adjacent meeting points.

buckyball A polyhedral molecule with pentagonal and hexagonal faces.

chromatic polynomial The number of ways of coloring a map with a given number of colors. If there are λ colors, then the number of ways is a **polynomial** in λ.

closed cycle A sequence of lines on a diagram that pass only once through any point and return to the starting point.

configuration A connected collection of countries in a map.

counting formula If C_k is the number of k-sided countries in a **cubic map**, then $4C_2 + 3C_3 + 2C_4 + C_5 - C_7 - 2C_8 - 3C_9 - 4C_{10} - \ldots = 12$.

country A region in a map.

C-reducible configuration A configuration that can be proved to be **reducible** only after it has been modified in some way.

cube A regular polyhedron bounded by six squares.

cubic map A map in which there are exactly three boundary lines at each meeting point.

cubic polyhedron A polyhedron in which exactly three faces meet at each vertex.

cuboctahedron A semiregular polyhedron with eight triangular and six square faces.

cycle See **closed cycle**.

digon A two-sided country in a map.

discharging See **method of discharging**.

dodecahedron A regular polyhedron bounded by twelve regular pentagons.

D-reducible configuration A configuration for which every coloring of the surrounding ring is a **good coloring** or can be converted into one by one or more **Kempe-chain changes** of color.

earth-moon problem Given two maps on the earth and the moon in which each earth-country has a moon-country as a colony, how many colors are needed to color their countries if each earth-country receives the same color as its moon-colony?

edge (of polyhedron) A line along which two faces of a polyhedron meet.

empire problem The problem of coloring a map with several empires, each consisting of a mother country and a number of colonies that must be colored the same as the mother country.

Euler's formula Short for **Euler's polyhedron formula**.

Euler's formula for the *h*-holed torus For any map drawn on an **h-holed torus**,
  (number of countries) − (number of boundary lines)
  + (number of meeting points) = 2 − 2*h*.
Euler's formula for maps on the plane or sphere For any map drawn on the
  plane or sphere,
  (number of countries) − (number of boundary lines)
  + (number of meeting points) = 2.
Euler's formula for the torus For any map drawn on the **torus**,
  (number of countries) − (number of boundary lines)
  + (number of meeting points) = 0.
Euler's polyhedron formula For any polyhedron,
  (number of faces) − (number of edges) + (number of vertices) = 2.
exterior region The infinite part of the plane lying outside the main part of a
  map.
face (of polyhedron) One of the flat surfaces that make up a polyhedron.
finitization of the four-color problem Reducing the proof of the four-color
  problem to investigating only a finite number of configurations.
five-color theorem Every map can be colored with at most five colors in such a
  way that neighboring countries are colored differently.
four-color problem Can every map drawn on the plane be colored with at most
  four colors in such a way that neighboring countries are colored differently?
four-color problem for a sphere Can every map drawn on the surface of a
  sphere be colored with at most four colors in such a way that neighboring
  countries are colored differently?
four-color theorem Every map drawn on the plane can be colored with at most
  four colors in such a way that neighboring countries are colored differently.
four colors suffice A statement of the four-color theorem.
four-legger country One of Heesch's three **obstacles to reducibility**.
geographically good configuration A configuration that contains neither of
  Heesch's first two **obstacles to reducibility**.
golden ratio The number ½(1 + 5)= 1.618034 . . .
good coloring A coloring of a ring of countries that can be extended directly to
  a coloring of the countries inside the ring.
graph See **linkage**.
great rhombicuboctahedron A semiregular polyhedron with twelve square,
  eight hexagonal, and six octagonal faces.
Guthrie's problem Another name for the four-color problem.
Hamiltonian cycle A sequence of lines on a diagram that pass exactly once
  through every point and return to the starting point.
hanging 5–5 pair One of Heesch's three **obstacles to reducibility**.
Heawood conjecture For each positive number *h*, there is a map on the surface
  of an **h-holed torus** that requires $H(h) = ½(7 + \sqrt{1 + 48h})$ colors; this conjecture was proved by Ringel and Youngs.
Heawood number The number $H(h) = ½(7 + \sqrt{1 + 48h})$ associated with the
  Heawood conjecture.
***h*-holed torus** A ring-shaped surface with *h* holes.
horseshoe See **Shimamoto horseshoe**.
icosahedron A regular polyhedron bounded by twenty equilateral triangles.

icosian calculus An algebraic system of symbols that satisfy certain equations; it was introduced by Hamilton.

Icosian Game A game in which Hamiltonian cycles are traced on a dodecahedron.

induction See **mathematical induction**.

Kempe chain Part of a colored map consisting of countries colored with just two colors.

Kempe-chain argument See **method of Kempe chains**.

Kempe-chain change of color An interchange of two colors in part of a colored map.

knot problem The problem of determining whether a given tangle of string in three dimensions contains a knot.

linkage, or **graph** A diagram obtained from a map, with points representing countries and linked points corresponding to neighboring countries. Coloring the map corresponds to lettering the points so that any points that are linked are lettered differently.

m-and-n rule A useful rule of thumb for identifying reducible configurations.

map A collection of countries or regions separated by boundary lines.

mathematical induction A method of mathematical proof. As applied to maps, one derives the truth of a statement for maps with $n + 1$ countries from the corresponding statement for maps with n countries and then deduces the result for all maps.

meeting point A point in a map where boundary lines and countries meet.

method of discharging A method for proving that a set of configurations is an **unavoidable set**: each k-sided country is assigned a "charge" of $6 - k$, and the charges are then moved around the map so that the total charge remains unchanged.

method of Kempe chains A method of coloring maps in which two colors are interchanged to enable the coloring of countries that could not previously be colored.

minimal counterexample See **minimal criminal**.

minimal criminal A map with a certain number of countries that cannot be colored with four (or any other given number of) colors, while any map with fewer countries can be so colored.

neighboring countries (or **regions**) Two countries (or regions) with a boundary line in common.

obstacles to reducibility Three arrangements of countries (a **three-legger articulation country**, a **four-legger country** and a **hanging 5–5 pair**) whose appearance in a configuration seems to prevent it from being a **reducible configuration**.

octahedron A regular polyhedron bounded by eight equilateral triangles.

"only five neighbors" theorem Every map (on the plane or sphere) has at least one country with five or fewer neighbors.

"only six neighbors" theorem for the torus Every map on the torus has at least one country with six or fewer neighbors.

pentagon A five-sided country in a map.

Petersen graph A diagram with ten points and fifteen lines that has no **Hamiltonian cycle** and does not arise from a polyhedron.

Platonic solid See **regular polyhedron**.

Poincaré conjecture A problem concerning spheres in four-dimensional space. It was proved by Grigory Perelman in 2002.

polyhedron A solid shape bounded by plane faces.

polynomial An expression, such as $\lambda^4 - 5\lambda^3 + 8\lambda^2 - 4\lambda$, consisting of multiples of powers of a variable, in this case λ.

problem of the five palaces Five sons each build a palace. How can the five palaces be linked in pairs by roads so that no two roads cross?

problem of the five princes A king in India had a large kingdom and five sons. How can his kingdom be divided into five regions so that each region has a boundary line in common with the remaining four regions?

quadratic algorithm for map coloring A map-coloring algorithm with running time proportional to n^2, where n is the number of countries

qualitative approach In the case of the four-color theorem, an attempt to prove it by showing that all maps of a certain type can be colored with four colors.

quantitative approach In the case of the four-color theorem, an attempt to prove it by finding the number of ways in which a map can be colored with any given number of colors.

quartic algorithm for map coloring A map-coloring algorithm with running time proportional to n^4, where n is the number of countries.

reducible configuration A configuration that cannot occur in a **minimal criminal**. If a map contains a reducible configuration, then any coloring of the rest of the map with four colors can be extended (possibly after some recoloring) to a coloring of the entire map.

reduction obstacles See **obstacles to reducibility**.

region A general term for a country, county, or state in a map.

regular polyhedron or **Platonic solid** A polyhedron in which all faces are regular polygons of the same type, and each vertex has the same arrangement of polygons around it. There are only five regular polyhedra—the tetrahedron, cube, octahedron, dodecahedron, and icosahedron.

ring of countries The countries immediately surrounding a given configuration. If there are k countries in the ring, the configuration is called a k-ring configuration.

ring size The number of countries surrounding a configuration. If there are k surrounding countries, the configuration has ring size k.

second neighborhood The second neighborhood of a country C is the collection of all those countries that are not themselves adjacent to C but are adjacent to a country that is adjacent to C.

semiregular polyhedron, or **Archimedean polyhedron** A polyhedron in which all the faces are regular polygons, not all of the same type, and each vertex has the same arrangement of polygons around it.

seven-color theorem for the torus Every map drawn on the torus can be colored with at most seven colors, and there are torus maps that need all seven colors.

Shimamoto horseshoe A configuration with **ring size** 14 whose **D-reducibility** would have implied the four-color theorem.

six-color theorem Every map can be colored with at most six colors in such a way that neighboring countries are colored differently.

square A four-sided country in a map.

stereographic projection A projection, from the North Pole, of a map of the surface of a sphere onto the plane on which the sphere sits.

tetrahedron A regular polyhedron bounded by four equilateral triangles.

three-legger articulation country One of Heesch's three **obstacles to reducibility**.

torus A ring-shaped surface with a hole.

triangle A three-sided country in a map.

truncated icosahedron A semiregular polyhedron with twelve pentagonal and twenty hexagonal faces.

truncated octahedron A semiregular polyhedron with six square and eight hexagonal faces.

unavoidable set of configurations A collection of configurations, at least one of which must appear in every map. The simplest example consists of a digon, a triangle, a square, and a pentagon.

vertex of polyhedron A corner of a polyhedron.

Albers, Donald, 157–58
algorithms, 167, 187
Allaire, Frank, 150, 152, 155, 158, 164
antiprism, 31
Appel, Andrew, Laurel, and Peter, 152
Appel, Kenneth, 3, 126, 130, 134, 139, 141–67, 173
April Fool, 10
Arago, François, 28
Archdeacon, Dan, 167
Archimedean polyhedron, 31, 187
Athenaeum, 18–19
Australian map, 67
A Voyage around the World, 81–82
axiom, 17

Babbage, Charles, 16
bagel, 20, 97, 100, 102
Baltzer, Richard 26, 72–73, 171
Bell, Eric Temple, 27
ben Gerson, Levi, 49
Berge, Claude, 124
Berman, Gerald, 123
Bernhart, Arthur, 115
Bernhart, Frank, 115, 151, 158
Bernhart, Mrs., 115
Bers, Lipman, 153–54
Birkhoff, George David, 105, 112–16, 120–22, 124, 131, 172
Birkhoff, Mrs., 115
Birkhoff diamond, 115–20, 131–32, 146, 187
Bishop of London, 73, 137
Boone, Bill, 130
boundary line, 3, 187
buckminsterfullerene, 32
buckyball, 32, 43, 91, 187
Busacker, Robert, 124

Californian traffic cop, 126
Carroll, Lewis, 71

Catherine the Great, 28
Cauchy, Augustin-Louis, 36, 58, 171
Cayley, Arthur, 45–49, 55, 57–58, 71, 88, 171
Cayley, Sir George, 45
cell of a bee, 80, 84
charge, 109, 139
chessboard map, 5
chromatic polynomial, 121–23, 187
closed cycle, 76, 79–80, 187
coefficients, of polynomial, 121
Cohen, Daniel, 162
Collingwood, Stuart, 71
coloring boundary lines, 75–77, 79
coloring maps, 1
coloring maps on honeymoon, 115
coloring maps on the torus, 20, 97–102, 125
commutative, 16–17
configuration, 107, 187
Cook Wilson, John, 73
Coq, 168
coterminous boundary lines, 76
counting formula, 40, 42, 187
country 3, 187
Courant, Richard, 126, 128
Coxeter, H.S.M., 27, 162
C-reducible configuration, 120, 131, 146, 168, 187
cube, 30, 33, 35–36, 38, 79, 187
cubic map, 8, 40–43, 75–77
cubic polyhedron, 43–44, 78–79, 82, 187
cuboctahedron, 31, 34, 187
cup of tea, 25
cycle, 76–77, 79–84, 187

David, 129
Delannoy, H., 102
de la Vallée Poussin, C., 92, 102

De Morgan, Augustus, 12–20, 45–47, 58, 116, 167, 171
Descartes, René, 34–35
digon, 39–40, 43, 52, 106, 108–9, 187
discharging, 108–12, 146–47, 187
dodecahedron, 30, 33, 36, 79, 81–83, 107, 187
Dodgson, Charles Lutwidge, 71
dominoes, 49–50
D-reducible configuration, 119–20, 131–32, 136–38, 168, 187
dual formulation, 134
dual graph, 67
Durham Castle, 87
Dürre, Karl, 131–38, 172

earth-moon problem, 97, 187
edge of polyhedron, 32, 187
Einstein's theory of relativity, 10
Ellis, Robert, 18, 171
empire problem, 95–97, 187
Errera, Alfred, 88, 91–92, 116, 120
Euclid's Elements, 30, 71
Euler, Leonhard, 28–29, 32–33, 35, 171
Euler's formula for the h-holed torus, 100–102, 125–26, 188
Euler's formula for the plane, 33–41, 58, 68, 96, 98, 111, 188
Euler's formula for the torus, 38, 98, 188
Euler's polyhedron formula, 33–34, 188
exterior region, 6–7, 40, 188

face of polyhedron, 32, 188
Feit, Walter, 161
Fermat's last theorem, 161
finitization of the four-color problem, 139, 188
"first cuckoo of Spring," 169
five-color theorem, 92–95, 188
four-color problem, 1, 7, 12–13, 129, 188
four colors suffice, 149–50, 188
four-color theorem, 2, 188
four-legger country, 140, 143, 188
Franklin, Philip, 105, 111, 115–16, 120, 124, 172
Frederick the Great, 28
fried chicken, 108
Fuller, Buckminster, 32

Galton, Francis, 19, 47
Gardner, Martin, 10, 160, 173
geographically good configuration, 142–45, 188
Gilman, Daniel, 56, 68
Gödel's incompleteness theorem, 130
Goldbach, Christian, 29, 32–33, 171
golden ratio, 122, 188
Goliath, 129
Gonthier, Georges, 168, 173
good coloring, 117–19, 155, 188
Gorenstein, Daniel, 161
Gounod, Charles, 56
graph, 67, 124, 134, 188
great rhombicuboctahedron, 31, 34, 44, 188
Guthrie, Francis, 3, 14, 19–20, 171
Guthrie, Frederick, 14–15, 19, 77, 171
Guthrie's problem, 14, 188

Haken, Armin, 152, 158
Haken, Dorothea, 149, 152
Haken, Wolfgang, 3, 108, 126, 128–31, 134, 137–67, 173
Hamilton, Sir William Rowan, 12–13, 15–17, 79, 81–83, 167, 171
Hamiltonian cycle, 82, 84–85, 103, 188
hanging 5–5 pair, 140–41, 144
Harary, Frank, 124
Hardy, G. H., 162
Heawood, Percy John, 66, 86–89, 91–92, 95–97, 100–104, 106, 112, 125, 172
Heawood conjecture, 102, 125–26, 167, 188
Heawood map, 51, 88–91
Heawood number, 101–2, 125, 188
Heesch, Heinrich, 108, 126–28, 130–32, 134–37, 139–41, 150, 155–56, 166, 172–73
Heffter, Lothar, 101–2, 125, 172
heptagon, 109–10, 134, 139, 142
Herschel, Sir John, 16
hexagon, 31–32, 34, 43–44, 106, 108–9, 111, 139, 142
h-holed torus, 100–2, 125, 189
Hilbert, David, 126–27
horseshoe, 137–39, 148, 157, 188–89
Hotspur, 18

icosahedron, 30, 34, 189
icosian calculus, 81, 189
Icosian Game, 82–83, 189
induction, 49–50, 189

Jackson, Brad, 96, 173
Jacques, John, 82–83
Johns Hopkins University, 56, 67–68, 71

Kainen, Paul, 164–65, 173
Kelvin, Lord (William Thomson), 73
Kempe, Alfred Bray, 55, 57–60, 65–69,
    71–72, 74–75, 77, 86–88, 92, 95, 102–3,
    105–7, 112–15, 124, 134, 161, 171
Kempe-chain argument, 60, 94, 106, 112,
    118–20, 155, 163–64, 189
Kirkman, Thomas Penyngton, 79–84, 171
Klein, Felix, 72
knot problem, 129–30, 189
Koch, John, 146, 148, 156, 158, 160, 173
Kohinoor diamond, 115
König, Dénes, 124
Kraken, 153

Laplace, Pierre-Simon, 29
Lebesgue, Henri, 111, 124, 172
Legendre, Adrien-Marie, 35, 171
Leibniz, Gottfried Wilhelm, 35
Leonardo da Vinci, 10
Levi ben Gerson, 49
Lewis, D. C., 120, 122, 172
Lhuilier, Simon-Antoine-Jean, 38, 98, 171
Liddell, Alice, 71
likely-to-be-reducible configuration, 141
linkage, 57, 66–67, 124, 134, 189
Listing, Johann Benedict, 21
logic, 21
Lucas, Edouard, 71–72, 92, 102

Maddison, Isabel, 27
m-and-n rule, 143, 189
Mansion, Pierre, 102
map, 3, 168, 189
mathematical induction, 49–50, 189
mathematical proof, 2–3
May, Kenneth, 2
Mayer, Jean, 150, 155, 163
McGregor, William, 10

McGregor's map, 11
meeting point, 3, 36–37, 189
method of discharging, 108–12, 146–47,
    187
method of Kempe chains, 60, 189
method of mathematical induction, 49–50,
    189
minimal counterexample, 51, 189
minimal criminal, 51, 92, 106–7, 124, 189
Minkowski, Hermann, 105
Möbius, August Ferdinand, 20–22, 72, 171
Möbius's problem of the princes, 20,
    22–27, 72, 97, 125, 190
Möbius's problem on a torus, 24
Möbius strip/band, 21
Moore, Edward F., 133, 163, 172

neighboring countries or regions, 3,
    23–24, 26, 110, 139, 189
Newcomb, Simon, 56
Norwegian sea monster, 152

obstacles to reducibility, 140–43, 189
octagon, 109–10, 134, 139, 142
octahedron, 30, 33, 189
Oedipus, 13
Oliver, J. E., 19
"only five neighbors" theorem, 39, 42, 59,
    106, 189
"only six neighbors" theorem for the
    torus, 98–99, 190
Ore, Oystein, 124, 134, 172
Ortoni, Andrew, 153
Osgood, Thomas, 142

parquet problem, 127
Pasta, John, 134
patch, 48, 59
Peaucellier's linkage, 57
Peirce, Benjamin, 19
Peirce, Charles Sanders, 19–20, 68–70, 171
pentagon, 30–32, 34, 39, 43, 60, 62–65, 89,
    93, 106–12, 116, 122, 134, 139, 141–42,
    190
Petersen, Julius, 102–3
Petersen graph, 103, 190
pie map, 5
Plato, 30

Platonic solids, 30, 190
Poincaré conjecture, 129–30, 190
polyhedron, 29–36, 38–40, 43–44, 190
polyhedron formula, 33–34, 188
polyhedron with tunnel, 38, 92
polynomial, 121–23, 190
prism, 31
probability, 104
problem of the five palaces, 22–23, 67, 190
problem of the five princes, 20, 22–27, 72, 97, 125, 190
projection, 7, 35–36
proof, 2–3
proof checker, 168

quadratic algorithm, 167, 187, 190
qualitative approach, 120, 190
quantitative approach, 120, 190
quartic algorithm, 167, 187, 190
quaternions, 16–17
Queen Victoria, 71

Ramsey, A. S., 102
reducible configuration, 105, 107, 112–15, 119–20, 131–32, 136–38, 146, 148, 151, 156, 160, 168, 187, 190
reduction obstacles, 140–43, 190
region, 3, 190
regular parquet problem, 127
regular polyhedron, 30, 190
Reynolds, Clarence, 116, 120, 172
Ringel, Gerhard, 96–97, 125–26, 172–73
Ringel, Mrs., 126
ring of countries, 113–15, 190
ring size, 132–34, 136, 139–40, 143–44, 146–47, 155
Robertson, Neil, 166–68, 173
Russell, Bertrand, 163

Saaty, Thomas L., 164–65, 173
Saeki, S., 165
Sanders, Daniel, 166, 173
Schmidt, Ulrich, 165, 173
second neighborhood, 166, 190
semiregular polyhedron, 31, 190
seven-color theorem for the torus, 99–100, 190
Seymour, Paul, 166, 173

Shakespeare, 18
Shimamoto, Yoshio, 135–38, 140–41, 159, 173
Shimamoto horseshoe, 137–39, 148, 157, 188–89
six-color theorem, 54, 191
Smith, Henry, 46, 88
soccer ball, 32, 43
solid angle, 32, 34
Spencer-Brown, George, 162–64
sphere, 6–7, 35
Sphynx, 13
square, 5, 29–31, 33–34, 39–40, 43–44, 52–53, 60–62, 93, 106, 108–9, 111–12, 191
Steinberger, John P., 168, 173
Stemple, Joel, 124, 172
stereographic projection, 7, 35–36
Stewart, Ian, 161
Story, William, 56, 67–70
Stromquist, Walter, 140, 144, 150, 168
Swart, Ted, 150, 155, 157–60
Sylvester, James Joseph, 45–46, 55–57, 67–68

Tait, Frederick Guthrie, 73
Tait, Peter Guthrie, 71, 73–75, 77–79, 82–83, 85, 102–3, 108, 124, 171
tangle, 129
Taylor, Herbert, 96
Temple, Frederick, 73, 171
tetrahedron, 30, 33, 37, 191
Thomas, Robin, 166–67, 173
Thompson, John, 161
Thomson, William (Lord Kelvin), 73
three-legger articulation country, 140, 143, 191
Tietze, Heinrich, 15, 20, 22
tiling problem, 127
torus, 20, 24, 97–102, 125–26, 191
traffic cop, 126
triangle, 29–31, 34, 39–40, 43, 52, 60, 106–9, 111, 191
truncated icosahedron, 31, 34, 43, 101
truncated octahedron, 31, 34, 43, 79, 191
Tutte, William T., 85, 122–23, 138, 140, 150, 152–53, 158, 168–69, 172
Tymoczko, Thomas, 159–60, 167, 173

unavoidable set, 105–9, 111–12, 126, 128,
    141–42, 152, 156, 159–60, 162–63, 166
unavoidable set of reducible configura-
    tions, 107, 126, 128, 139, 141, 144, 148,
    152, 159, 166
U.S. map, 4–6

Valéry, Paul, 116
Veblen, Oswald, 105, 112
vertex of polyhedron, 32–34, 191
voyage around the world, 81–82

wee planetoid, 84
Weise, Karl Heinrich, 129

Weiske, Benjamin Gotthold, 21, 72
Wernicke, Paul, 107–8, 124, 172
Weyl, Hermann, 126
Whewell, William, 17–18, 47, 171
Whitney, Hassler, 105, 121, 124, 138, 140, 172
Wiener, Norbert, 111
Wiles, Andrew, 161
Wilson, James Maurice, 72–73
Wilson, Robin, 47, 163, 167
Winn, C. E., 116, 120, 172
Witt, Ernst, 128
Wright, Chauncey, 19

Youngs, Ted, 125–26, 172